THE LAST
MUSTER

CARLY THOMAS

THE LAST MUSTER

HarperCollins*Publishers*

HarperCollins*Publishers*
Australia • Brazil • Canada • France • Germany • Holland • India
Italy • Japan • Mexico • New Zealand • Poland • Spain • Sweden
Switzerland • United Kingdom • United States of America

First published in 2024
by HarperCollins*Publishers* (New Zealand) Limited
Unit D1, 63 Apollo Drive, Rosedale, Auckland 0632, New Zealand
harpercollins.co.nz

Copyright © Carly Thomas 2024

Carly Thomas asserts the moral right to be identified as the author of this work. This work is copyright. All rights reserved. No part of this publication may be reproduced, copied, scanned, stored in a retrieval system, recorded, or transmitted, in any form or by any means, without the prior written permission of the publisher. Without limiting the author's and publisher's exclusive rights, any unauthorised use of this publication to train generative artificial intelligence (AI) technologies is expressly prohibited.

A catalogue record for this book is available from the National Library of New Zealand

ISBN 978 1 7755 4 2285 (paperback)
ISBN 978 1 7754 9 2597 (ebook)

Cover design by Louisa Maggio, HarperCollins Design Studio
Jacket images: Front cover, back cover and back flap photos by Francine Boer; front flap photo by Carly Thomas; author photo by Ava Thomas
Endpapers: photos by Francine Boer
Typeset in Adobe Caslon Pro by Kirby Jones
Printed and bound in Australia by McPherson's Printing Group

To those who turn to a gentler way

and know the worth of a good horse.

Catching the horses in the early morning, Middlehurst Station.
Francine Boer

CONTENTS

Introduction		1
1.	The last long muster – Greenstone Station	9
2.	High-country horses – Muller Station	49
3.	The hard stuff behind – Awapiri Station	79
4.	Last ones standing – Tiroroa Station	113
5.	In the blood – Ruanui Station	129
6.	Pretty magic – Smedley Station	149
7.	The real deal – Mangaheia Station	171
8.	Survival of the fittest – Ngahiwi & Awapapa Stations	195
9.	Anything is possible – Middlehurst Station	217
10.	Wild horses – St James Station	247
11.	Free rein – Bluff Station	265
Acknowledgements		295

INTRODUCTION

Some years ago, I went on a muster at Papanui Station, in the green blocks outside of Taihape, in the central North Island. That day, after saddling our horses and riding cloud-bound in a clinging mist, we paused and, for a moment, all that could be heard was the crunching of the horses on their bits, the huff and puff of their dragon breath, the whine of an eager dog. The morning was still, calm, enveloping – and then a mechanical sound broke the peace. A quad bike roaring in the distance, over the valley on the opposing hill. Papanui's neighbour, busy mustering his block his way.

As the bike came into view, it halted abruptly at the base of a straight-up-and-down hill. The neighbouring farmer started throwing angry shouts at his cattle, but they simply stared at him, hard-nosed and unbudging.

When the hills get high, horses go where machinery can't. Francine Boer

Introduction

I looked over at Papanui's manager, Guy, just in time to catch the wry smile forming on his face. 'And that's why we use horses here,' he said calmly, then gave a whistle command to his heading dog.

On that muster, I got to see the true worth of a horse. We were doing things the old way, on horses bred for the terrain. It was slower – we waited, we watched the movement of the sheep – but we could also sidle in to places that could never have been reached on a bike. And we still got the job done efficiently. The mob was in the yards right on time. What's more, the sheep were worked without stress.

Horses have been entrenched in Aotearoa's culture since 1814, when missionary Samuel Marsden first brought them to our lands. When these creatures – two mares and a stallion – were swum ashore, local iwi Ngāpuhi thought they were taniwha, then watched, awe-filled, as the majestic long-legged beasts strode across the sand. Horses were quickly assimilated into Māori lives, and often gifted to local chiefs by government officials. By 1840, Māori were breeding their own hōiho, and outnumbered Pākehā in terms of horse ownership. Tūhoe used them to travel into Te Urewera with a network of tracks, and Ngāti Porou also used them for travelling and hunting in the East Coast.

Horses were kings on the land for European settlers, too. Every farm in the mid-1800s had at least one because, quite simply, there was no other option to muster stock. Apart from walking – but, as I have heard it said many times over, 'Why walk when you can ride?' By 1911, there were 404,284 horses in New Zealand – hefty Scottish Clydesdales flexed their muscles, pulling ploughs to work the newly cleared flatlands, and in the hills Hackneys, thoroughbred crosses and standardbreds were among the first breeds mustering stock and carting goods as packhorses. Our farms and stations were built on the backs of these horses, and these sturdy beasts even shaped much of our early infrastructure – many of the roads we travel on today are old coaching ways, and early colonial settlements were set up to be within rideable distances of one another.

But, in 1910, tractors arrived, and so began the downward tumble of our equine workforce. Horses were still needed for the hills where tractors couldn't go, but machinery was seen as new, progress, the way of the future. Heavy ploughing harnesses and strapping were hung up, one by one, and by the middle of the century most were little more than museum pieces. And then quad bikes came along in the 1970s, further hastening

Station horses are bred for purpose. They are strong, agile and brave.
Ava Thomas

Mustering on horseback makes for some magic moments.

the equine decline. New Zealanders were fast to make the transition to bikes, and we're now among the highest users in the world.

In 2022, just over 33,500 horses were still being used on New Zealand farms – a decrease over the preceding decade of more than 20,000. Horses are an increasingly rare sight in our hills. And it's not just tractors and quad bikes to blame; they've been further phased out by helicopters and drones, too. What's more, farming in New Zealand is in a transition stage – pastoral properties are being converted to intensive farms and forestry operations, and there are ever-tighter rules, complex processes and an increasing emphasis on environmental concerns. The traditional landscape is changing. As pressures on farmers mount, the old ways are being pushed into the margins by methods perceived to be faster and more economical.

But you can't pat a motorbike, and they won't be your friend. And, after that day well spent on Papanui Station, I had to ask: Do horses really need to be a thing of the past? Or could they still be relevant and viable? Could a step back to horses in fact be a step forward? Could horses perhaps help us to navigate these uncertain times of increasingly unpredictable and severe weather?

Introduction

So many questions, and I wanted answers. So I went in search of horses and yarns, history and hard yakka, something old and passion-filled. Ride with the best of them, that was the grand plan, all while hearing the reasons these folk hold the reins so tightly on their way of life.

1
THE LAST LONG MUSTER

Greenstone Station, Glenorchy, Otago

Stu Percy's face unfolded in a wide smile when he saw the convoy of dusty utes and horse floats rolling down the long driveway of Greenstone Station. Autumn is a busy time in the South Island's high-country farming calendar; when the leaves start to let go of their summer hold, it's time to muster stock closer to home. Time for weaning, for getting things done before the harsh winter closes in, and a chance for the Greenstone manager to catch up with old friends.

The musterers clambered out. Doors slamming, scuffed boots crunching the gravel, weathered coats, favourite jeans faded to light blue, shirt collars turned up against the cold. Greeting one another with handshakes, backslaps and jokes picked up from where they left off. Stu's wife, Anne, had assured me new faces were always welcomed, and I was soon wrapped into the fold of musterers.

The Percys had agreed to let me tag along on the Greenstone autumn muster a month earlier, and ever since I'd been poring over the route, tracing the ridges and contour lines of the epic 42-kilometre, five-day journey into the Greenstone Valley and back out again. This was the shape the muster had always taken, with many parts of the track remnant of the way followed by musterers over a century earlier. It was a piece of history and it was a privilege to be a part of it.

We were a stone's throw from the western shores of Whakatipu Waimāori/Lake Wakatipu, and under the blue sky a chill breeze was blowing. Horses in the wooden-railed yard nodded their heads towards me, and as I walked over to them – halter in hand and instructions to catch the tallest – a handsome dark bay, bolder and bigger than the rest, came to nudge my hand. This friendly gelding was definitely fitting the description. Sergeant was his name and he had a good dose of Clydesdale running in his blood.

Central Otago was one of the first places in New Zealand to import Clydesdales, which originated from a fifteenth-century ancient Scottish lowlands breed. The early horses that set their dinner-plate-sized hoofs on our soil came from the Valley of Clyde, where English and Flemish blood had been introduced. They were large, with short, muscular backs, well suited to pulling the equipment needed to break in the land for cropping. A beautiful characteristic of the Clydesdales were their 'feathers' – wispy, long-falling hair that covered their fetlock joints – and Sergeant sported a fine set of feathers. I could see that it was not going to be an easy job getting on him – I had to get up on the railing just to be able to slip the bridle over his teddy-bear ears – but he was a gentleman and stood quietly the whole time. Used as a trekking horse before he came out to Greenstone, he thankfully had more whoa than go.

As I readied myself and Sergeant, I soaked in my surroundings. On the rail nearby, dusty and well-worn saddles were lined up, patiently waiting for similarly well-worn saddle bags to be fixed in place, and wet-weather gear tied on with blue baling twine – the sort usually used to secure hay bales. It's the fix-all necessity of rural New Zealand, and all good shepherds carry a piece in every pocket of every coat. Then there were the horses: standardbreds like Stu's steed, Sergeant the heavy Clydesdale cross, and still others with thoroughbred, warmblood and Arabian bloodlines – horses that represented the earliest breeds introduced to New Zealand, back in the 1800s. This dusty post-and-rail yard at Greenstone was a melting pot of New Zealand's equine history.

There was a real hum of adventure and at last, everyone was ready. All nine of us mounted up and moved out, led by Stu, with 31 dogs running at our horses' heels. I threw a goodbye wave to Anne, who was staying behind to hold the home fort, then we rode along Elfin Bay with its white stones rounded by the lake's cold waters. The energy was big, boisterous and generous, and I was totally swept up in it. But, mingled with all that excitement was a heavier note. For this was more than just another autumn muster. It was to be the station's last. The end of an era.

*

When Stu and Anne Percy first came out to manage Greenstone Station, it was 'just for a few years'. That was back in 1982. The couple were newly married, and the road from Queenstown was still gravel. Furthermore, the 21-kilometre stretch from Kinloch down Lake Wakatipu's western shore had only just been put in. Nowadays, the drive from Queenstown takes just over an hour, but it was easily double that back then, when the road wasn't much more than a dirt track and there were 26 fords to navigate. 'You never got out of second gear,' remembered Stu. 'We got stuck all the time. The road was marginal, often impassable.'

It was easier to bring things in over the water ... sort of. 'You try lifting a big old washing machine into a jet boat!' Stu said. Still, a jet boat was an improvement in terms of speed; in earlier times, the station's inhabitants relied on the grand TSS *Earnslaw* steamboat for the carriage of goods, while a barge was used to carry livestock across the lake to the wharf at Queenstown.

Back in the eighties, the station was 10,000 acres, and owned by an American multi-property owner who changed its name to Greenstone. Prior to that, it had been known as Birchdale, with its first European runholder turning up to farm the land in the 1880s. He'd described the station homestead as a 'commodious dwelling', but his young wife had clearly thought otherwise – she quickly persuaded him to sell up and move back to 'civilisation'. To this day, that old homestead still stands in the valley, on a high terrace above the Caples River, enjoying a slow decline back to nature.

Heading out with all the excitement of a first day under clear skies.

Stu Percy is the laid-back lead of the muster and the manager of Greenstone Station.

The living was still pretty basic when the Percys arrived a century later, with no electricity yet hooked up to the homestead and access to the station dictated by what the clouds were doing. If it rained – which it did, often – all those rivers, creeks and fords came up, and the Percys hunkered down. According to Stu, this was a good way to start off a marriage: biffed in at the deep end with nowhere to run from an argument. The couple went on to have three daughters (Briley, Michelle and Grace) and a son (Scott); the three eldest were homeschooled for the first five years, so it was hard graft. But the Percys loved it from the get-go. 'There is just so much space out here. Room to breathe,' says Stu. 'It's pretty special.'

Then, when they were ten years into the job, a treaty settlement with South Island iwi Ngāi Tahu meant that Greenstone Station was tied together with the two other stations at the head of the lake. It was a hard-won restoration for the iwi, who were again kaitiaki whenua of a small but significant part of the high country they'd once owned. The mountaintops in this part of Te Waipounamu are known as Kā Whenua Roimata (The Lands of Tears), in recognition of the suffering of the Ngāi Tahu people and, as kaumatua Tā Tipene O'Regan has explained, there is good reason for the name: in total, 11 high-country properties were agreed on to be passed over to the iwi once the Crown procured them on the open market, but only three were *actually* given back – Elfin Bay, Greenstone and Routeburn stations. Kā Whenua Roimata is a poignant reminder of a difficult past.

As part of the treaty settlement process, the mobs of sheep dwelling on the back blocks of Mount Bonpland, Death Valley and Scott's Basin were gathered up, and the high-altitude country was retired. And the Percys moved down the road, to the old Routeburn Station homestead, with a huge land expanse to be getting on with: an epic 36,000 hectares fell into their management. They got through those early years with the help of Ngāi Tahu's then general manager of property development, Ian McNabb.

The terrain on all three stations was mountainous and hard going. No four-wheel-drive tracks had ever been made in the hills – it was just too challenging and didn't make economic sense. This was not a place for vehicles; it was horse country.

And that suited Stu just fine. 'I'd rather ride a horse than walk,' he said. 'It's a great way to work.'

*

Of course, there's another way into these mountains: on your own two legs. It was Waitaha, one of the earliest groups of Māori in the area, who established routes through the Greenstone and Hollyford valleys as the easiest access between the West Coast and Central Otago. Kāti Māmoe, Ngāi Tahu and Ngāti Wairangi also used the routes, and the footfalls of tangata whenua were the first human explorations of this immense place. The Ngāi Tahu heritage of the whenua was linked into its creation mythology, and this potent history was fuelled by the seams of pounamu that ran through the region. This highly valued stone held mana, and in turn so did the land it rested in.

In more recent times, Ngāi Tahu have leased a large proportion of their land to the Department of Conservation (DOC) in perpetuity, and these areas, with their vast network of tramping tracks and huts – all maintained by DOC – are well used by trampers and hunters alike. The Greenstone autumn muster made use of sections of both the Greenstone and Caples tracks, which linked into two of DOC's Great Walks – the Routeburn and the Milford tracks – and, while they were indeed a step above your average New Zealand tramping scramble, they were still hard going in places. And those were the fancy bits of the

The Rat's Nest Hut is a bush oasis at the end of a long day's ride.

muster route. As Stu pointed out to me, as we set off on that crisp autumn morning, 'Horses are a better way to travel out here with all the river crossings. We need the horses on this terrain. They are our way in. And, like I always say, a second-class ride is better than a first-class walk.'

Horses had been used on this land since at least the 1800s, ploughing the lower paddocks, mustering stock and carrying shepherds out to the back blocks. In its earliest days, the station probably had its own stallion; every farm and station back then had horses, and would breed their own for what they needed. In war time, some even bred military horses to supplement their income. Around the same time, horses also played an important role in the first journeys made by Europeans into Greenstone Valley, then were used in successive years once a pack route was established – back then, the only land route through to the West Coast.

In more recent times, as newer and shinier options had become available, some might have been tempted to change gear and use the sky here at Greenstone instead of the land, employing helicopters to drop in stores and do the odd recce. Not the Percys. 'Way too expensive,' Stu said. 'And just not our style.' Instead, he swore by his standardbred horses for mustering work. Lean, straight-backed, long-bodied and strong, the breed in New Zealand can be traced back to the 1880s, when they were imported from Australia and the United States for harness racing by Ōtautahi/Christchurch businessman Robert Wilkin. The breed gets its name from the standard of covering a mile in two minutes 30 seconds, established as a standardbred qualification in 1879. These horses could indeed cover the ground, and they tended to be hardier and more level-headed than their posher counterpart, the thoroughbred. All this made them a good choice in the early days of mustering in New Zealand, especially in the vast South Island, where drovers had to travel long distances to pick up their mobs of cattle or sheep. Additionally, standardbreds were an economical choice, as they were often given away once their track careers were done and dusted.

For Stu, his standardbred herd was a cheap-as-chips way to get the job done well. The only downside: his freebie horses sometimes came with a few issues. But Stu sorted them out by putting them to work: 'Give them a job and they have to think,' he said.

Now, though, the time of the big autumn muster on Greenstone Station was coming to a necessary end. Ngāi Tahu was committed to making environmental improvements in the valley, and that meant keeping stock away from the water. As Stu explained, the many miles of waterways on Greenstone simply couldn't be fenced. 'It would ruin it here,' he said simply. 'It really would.'

And while he was sad to see the cattle go, to be the one closing the chapter on the muster, he also understood. 'I'm a fisherman, and I have grandkids. There are places in this country where you could once swim and now you can't. Something has to happen.'

*

With Lake Wakatipu at our backs, we started to climb up into stands of southern and silver beech. We had a six-hour ride ahead of us, to Bush Creek Hut, crouched below Mount Mavora and surrounded by the Thomson and Livingstone mountain ranges. As we climbed, our horses settled into their job, trundling on in a gentler rhythm while the air got noticeably cooler the further we strayed from the sun's reach.

Up front, Stu rode with his daughters, both of whom were still closely involved with the station where they'd grown up: Michelle and her husband, Andrew, lived and worked at Greenstone, as did Grace and her boyfriend, Mark. A competitive showjumper, Grace was every bit as comfortable on her flashy chestnut as her dad was on his standardbred, and Michelle might have been slight atop her heavy-set horse but she was absolutely at ease.

As the track opened up before us, the going became increasingly rocky. We wove through windfall trees, ducked our heads, held onto hats, and were soon skirting alongside Lake Rere, where the small and smooth-leaved mountain beech trees admired themselves in the still water, then heading deeper into our journey, the smell of horse sweat mingling with the crispness of the bush. At one point, the horses carried us over a narrow path of scree formed on the shoulder of a dramatic slip that had partially blocked the river. As the horses navigated the precarious terrain, their senses became more alert, ears flicking back and forth, steps shortened, nostrils snorting, checking the air. In tune with their surroundings, noticing things we didn't. Another big spread of scree saw us clear of the trees, and crossing a river where on one muster, I was told, the water had been so high the cows were forced to swim across.

'That was a tough one,' remembered Warren, a bull breeder from Southland and old friend of Stu's who had been a reliable part of this muster for donkey's years. He was up on a mare he'd bred himself, and paused to lean down

Bill is quick with a yarn and can cook up a storm over an open fire.

Pushing cattle up the Greenstone Valley in startling bright sunlight.

and cinch the saddle's girth a bit tighter; after a few hours of riding, his horse had become a girth hole leaner. 'The dogs were getting washed into the cows from the force of the river flow. The horses had to work hard that day. The river changes all the time,' Warren told me. 'You have to trust your horses on these crossings.'

Our horses were much more than a mode of transport out here, I was beginning to understand; they were also our source of safety.

*

When we'd been riding for about six hours, I heard one of the guys calling back to me. 'Carly! How's your butt doing?' It was Grimmer, a Glenorchy local who had shot straight out to the front on his standardbred when we'd left that morning, and in the hours since had demonstrated a finely honed knack for winding people up.

'It's numb,' I yelled back, grinning. Several hours in a saddle will do that to you, but it in no way dulled my enjoyment. Plus, I knew Bush Creek Hut wasn't too far away, just a bit further south, and the horses seemed to know it too, lengthening their strides as we drew closer.

The Greenstone River had remained our constant companion along the flats. Creeks veined into it from the surrounding hills and, in parts, the ground we rode over was boggy from springs. The tussocks here were bigger than anything I had ever seen. 'A place to lose cows,' more than one of my fellow musterers told me.

At last, Bush Creek Hut came into view up ahead. Though it stoically hunched in place, I'm pretty sure a swift wind could have blown it over. Like a child's drawing, it was all chimney and roof, added to and patched up over the years, but still with its original backbone of bare native timber. Character in bucketloads, enough firewood to keep us alive, some basic equipment, and a long-drop way up the back with a to-die-for view: it was all we needed.

'There are also whiskey bottles buried outside,' Grimmer told me with a wink. 'Left by hunters as a thank you. Stu knows all the hidey-holes.'

We tied our sweat-covered horses to old hitching rails made out of tree trunks, and took their gear off. Saddles perched on the high wood pile and bridles hung off rusty nails. Steam rose from our horses when we washed them down with water collected in buckets from the fast-flowing creek nearby, then they headed off to find overnight shelter in the folds of the hill above the hut. It was a scene that had been set many times, not just in this valley but all over New Zealand's big stations. By the late 1900s, huts just like Bush Creek had been built throughout the high country. And, since their dimensions were influenced by what building materials and equipment could be brought in with packhorses, they were mostly of the swing-a-cat variety. Literal shelters from the storm for shepherds, hunters and trappers.

Inside, one half of the hut was taken up with bunks, while at the other end a huge fire multitasked as a heater, open oven and boot dryer. The latter quickly became the domain of Bill, our sprightly camp cook, who in his 70-plus years had racked up many a muster and many a tall tale. Stoking the fire was an art form, and Bill was the master, growling at anyone who dared to place a log in not quite the right way. 'You buggers just stay away from it,' he said with a pointed finger.

We did as we were told, and settled in for a knees-up, crammed together, our sleeping spots claimed and all corners filled. As dark fell, a Bluetooth speaker blasting the Eagles and Katy Perry appeared, whiskey was shared around, and a feed of whitebait fritters was gratefully gobbled up. At some point, sleep found us. Well, some of us. The snoring that eventually set up inside that hut in the wee small hours was quite exceptional.

*

'I suppose we'd better go and find some cows,' said Warren the next morning, striped long-john-clad legs dangling over the edge of the bunk, one sock on and one sock off. Warren had the best-looking gear on the muster, and his horse was brushed to a high shine, but this morning even he looked a bit worse for wear.

Bill had already stoked the fire, and was soon serving up bacon, sausages, eggs, toast and beans that were enjoyed over good-humoured jibes about who'd done and said what the night before. 'Everything happens in its own time on this muster,' Stu said to me in his characteristically low-key manner, his hair sticking up on end. He was definitely looking a little dusty, too … but, if anyone was hungover this morning, they were sure as heck trying to hide it. Cattle were waiting to be found, and mustering them was what we were there for, after all. It took more than one coffee to get each of us out the door, but once the caffeine kicked in, the age-old art of saddling up commenced: a quick brush around the horses' saddle and girth areas, then the heave-ho to plonk weighty stock saddles on their backs.

I was delegated to take one side of the tussocked valley with Andrew, while two of the others took the opposite side. Meanwhile, the rest of the crew set off, the dogs with them bounding across the ground, and were soon at work on the valley floor, swiftly gathering a mob of cows with calves in tow. It was still cold up this high, and the cattle were scattered across the plain like boulders fallen from the mountains. Our task: find all the cows and get them into a forward-moving mob, an entity unto itself that mooed ever louder as its numbers increased.

From amid the huge tussocks, Andrew and I flushed out one cow and her calf, then pushed both into the forming fist of cattle. 'Here, take my horse,' Andrew said, passing me the reins and jumping to the ground. 'I'm going to see if there are any more hiding up in the hills.' Then he set off on foot, and soon disappeared from sight. Mustering isn't always all about the horses; there's also plenty of time spent on foot, clambering up steep inclines.

Andrew's little palomino quarter horse walked happily beside me and Sergeant, and together we continued pushing

Stu in the coveted front-of-the-fire spot at Bush Creek Hut.

The open expanse of the Greenstone Valley is a contrast to the surrounding hills.

the mob forward, enjoying the morning sun. I looked out over the tussock that had swallowed Andrew, and there wasn't a cow in sight. I was sure that area was clear. But then he reappeared, jumper off and tied round his waist, and trailing ahead of him were a dozen cows. Unlike me, Andrew knew where to look. And, with a similarly calm disposition to his father-in-law, he had just quietly got on with it and got the job done.

Once everyone's cows had come together in one large mob, we stopped for a late lunch, which we ate sprawled in the crunchy-dry grass, watching one last cow languorously make her way through the gate to join the rest. 'This is the life, eh?' said the softly spoken Ynyr. He was Stu's brother-in-law, and had a stockman's patience and serious character. Honestly, he could have ridden right out of the past century in his heavy-knitted jumper with leather patches on the elbows, thick jeans, leather chaps and mountain-tackling boots. His hat, horse and dogs were an extension of his own body.

'Yep,' replied Hayden, who was now in his thirties but had been joining the autumn muster since his late teens. He too had his hat low, and his words were slow and relaxed in the warm sun. 'This is what I wait all year for.'

The horses were similarly mellow, their heads dipped and their tails swishing lazily.

'Have you read the poems of Jim Morris?' Ynyr asked me suddenly.

'I can't say I have,' I replied.

Then Ynyr cleared his throat, sat up a bit straighter, and began to recite:

When the last long muster's over,
And it's time for me to rest,
I'll take this weary body
To the place I love the best.
To hear the river roaring
In a pent-up mountain flood,
And the wild Nor'west howling,
You get it in the blood.

He sank back onto the grass and pulled his hat back down over his face.

'Thanks, Ynyr,' I said. 'That was beautiful.'

He just grunted, but the echo of the poem hung in the autumn air.

The Last Muster

*

When we straggled back to Bush Creek Hut, it was to find a very tidy camp and Bill already stoking the fire. Our cook was in good spirits, pottering about while a big joint of home-kill and some potatoes cooked slowly in the embers of the fire.

I sat down next to him, hot cuppa in hand, and he got yarning about his past. In his heyday, he informed me, he drove cattle for a number of years. Before trucks took over, droving was an important aspect of New Zealand farming. 'On the hoof' it was called – taking stock from farms to the freezing works, saleyards, railheads and other farms. Drovers on horseback were a common sight on the roads, making their way with long strings of cattle and sheep, and there was a real skill to negotiating these large mobs over many kilometres.

'I took a mob from Ōpōtiki to Tuakau once,' Bill told me. (That, in case you're wondering, is a nigh-on 300-kilometre journey.) When he was asked to do the job, he'd had no horse and one dog. 'So I borrowed a horse and a couple more dogs, and that was enough to get me going.' But then there was the issue of the route: he didn't exactly know *where* he was going. He had no instructions, 'no nothing' (this said with a laugh and a slap of his cap on his leg), but he knew he needed to get the 600 cows moving. 'I had to organise paddocks up ahead, the works. It was a lot to sort out.' All up, it took him about six months to make it to Tuakau. On the way, he broke in two horses, ended up with 14 dogs, and got to know every pub along the route. 'I was ready for business,' he said with a low whistle.

By this time, the cold outside had started to set in and the others had begun to move indoors to join me and Bill for a feed of fire-roasted mutton.

'You talking horses again, Bill?' asked Stu.

'Of course,' Bill replied, grinning.

'Remember Jethro, Dad?' Grace chipped in. 'Now *there* was a horse.'

'Jethro was a neat horse,' agreed Stu, who was leaning precariously on a chair with a wobbly leg. 'Big standardbred. Taught all the kids how to ride. He was a cracker horse.'

'Actually, he got stuck in a bog one year and had to be dug out,' Grimmer chimed from one of the lower bunks, and Stu and Grace both nodded. 'A legend of a horse,' Grimmer went on. 'You could put anyone on him and go anywhere. Even someone who was drunk as would be fine.' Then he raised his cup. 'To Jethro!'

'To Jethro!' we all cheered.

'And what about Skipper?' Grimmer said, prompting a chorus of 'Skipper!' and more raised cups.

'Skipper was a packhorse,' Stu said to me while the commotion died down. 'He'd always chuck your load off, whether in the pen or all the way out here. Always wanted to get it off his back. He was a solid type, half Appaloosa, with a white spot in his eye. A real character. Just one of those horses. Got lost in the bush for a whole week once.'

'He could be a bugger,' Warren piped up, and was met with nods all round. 'There was one time when he had been fine all the way in … and then he wasn't.' Just as the musterers were getting to the hut after a long ride, Skipper had apparently decided he'd had enough. His thoughts turned to home, and off he went, taking the other packhorse – and all the team's supplies – with him. 'I came in later on that one,' said Warren with a grin. 'Saw Stu, and he didn't even say hello, which wasn't like Stu at all. He just said grumpily, "What have you got?"'

Stu chuckled. 'That's right. And you had brought in pāua, blue cod and whiskey, so we were all very happy to see you after that!'

Warren is a Southern bull breeder who never misses a muster if he can help it.

The vast expanse of tussock grass on the valley floor hides cows and bogs.

The night carried on with more talk of horses and good-hearted reminiscing, and at one point Stu turned to me and said, 'This muster has always been a good time. Just a really good catch-up. It's sad that we won't be doing this anymore.'

*

The next morning, the horses were way up the hill, as far away as they could climb. Horses have a tendency to position themselves in the most difficult place to be caught, and these fellas were no exception.

That night, we would be staying at another hut, so things were stuffed to overflowing in saddle bags or tied on with that trusty blue baling twine, while Stu and Bill – who would come along later with a packhorse carrying our food – started to clean up before moving out. Collectively, we attended to the unofficial but tightly adhered-to high-country hut code: whatever wood you burn gets replaced, whatever things you break you fix, and you leave the hut a little better than when you arrived.

Mounted up and down on the valley floor, we split up. 'Coming with me, Carly?' asked Grimmer, and I nodded, then followed him up towards the Mavora walkway, where there were yet more cattle to gather. Grimmer's dogs made fast work of it, and his huge huntaway forgot his sore paws as soon as he was asked to work. The cows gave us very little trouble – the only time they put up any fight was when they lost patience with the dogs, defending their calves by pushing into the barking pack with their heads down and hoofs flying. For a lumbering beast, they could move surprisingly fast, and at times we had to do some swift manoeuvring on our horses to bring them back into line. But, aside from these bursts of excitement, the task was one requiring patience: hours piled onto hours, as we slowly gathered and pushed, onwards and forwards, one hoof in front of the other.

We stopped for lunch – hot venison sandwiches prepared on a fire by Bill – at Pond Burn Hut, which was embedded in a forest of silver beech. We stretched cramped legs, wiggled a bit of life into numb bums and shook out cold fingertips. The horses and the dogs got to have a rest, too – the former nodded off as soon as they were tied to a tree, and the latter took the opportunity to curl into tight snoozing balls.

Girths were eased off, saddlebags rebalanced and hoofs checked. 'Bugger,' Warren muttered. 'Lost a shoe.' Then he set about shaping up a new one with a hammer, using the flat of a cut-down tree trunk as an anvil. In this rough terrain, shoes were absolutely necessary to keep the horses from going lame – but they were also inevitably pulled off on rocks. For this reason, extra shoes were carried, as well as the tools and the skills it took to refit one.

*

We reached Passburn Hut right on the golden hour – that time when day starts turning its face towards night, but delivers one final push of light that brings the landscape suddenly alive in a whole new way. We'd stopped at this hut for lunch on our first day. It sat just outside of the conservation area, above the Greenstone River, with Tooth Peak on its right shoulder and David Peaks on its left. Owned by Ngāi Tahu, it was used by musterers, hunters and trappers alike, and it was old – 'at least a hundred years,' Stu told me. Though not as old as its predecessor, dubbed Jock's Hut, which had stood in the exact same spot. The chimney that Jock built remained, a massive stone monument to days gone by, and Passburn had been built around it.

The moment we arrived, talk turned to who was going to go further up the valley to Rat's Nest Hut. Also known as Stoat's Nest, it was up near the station's northernmost boundary, with Milford lying just over the hills. Yet another bolt-hole for musterers, it was where calf-marking was done later in the year and it held the remnants of ancient yards. And there were cows up there that needed to be mustered, but it was an extra few hours' ride. I immediately put my hand up to go with Stu and Warren – I liked the sound of sleep with just two snorers, rather than eight – and the three of us were soon moving out again. Sergeant was less than impressed, giving an irritated swish of his tail before resigning himself to plodding on.

The temperature dropped dramatically, and soon the mountains had become the domain of shadows, ink-dipped origami folds of hard-edged rock. The river gorged deeply into swirling pools, its banks mingling with native bush – lush green ferns and spiky horopito. The light faded and disappeared just as we reached Rat's Nest, which sat with David Peaks at its back. It was made out of rusty old tin, with a propped-up crooked chimney and a lean-to that had leant so far it was on its knees. The whole construction was struggling to stay upright, but this Kiwi battler was stoical with its stout native-timber beams. Every hut in the valley was situated beside a creek, and Rat's Nest neighboured

Steele Creek, which tumbled into a waterfall that bubbled softly underneath the bird calls.

We found comfort in a fire, water and food, and as we sat enjoying our dinner I gazed at the graffiti on the hut walls. Each muster the Percys had done was recorded there, with all the names of the musterers listed. I recognised the names of this year's crew etched more than once, and couldn't help thinking how this living record was about to become a memorial. Yet another relic reminding visitors of the way things used to be done.

*

My peaceful snore-free slumber was broken early by a cry from Warren. 'There's cows right there!' he was saying. 'And they're heading the wrong way!'

Still half asleep, I got my feet out of my sleeping bag and on to the ground.

'They must have pushed through the gate,' said Warren.

'Bugger,' I replied sleepily, forcing myself to stand and start packing up and putting myself together. I soon learnt that Stu had already headed further up the valley to scout for cows, so it was now our job to hurriedly get behind the cattle and stop them from going any further in the wrong direction.

With a bit of speed from the horses and a lot of noise from ourselves and our dogs, we managed to get them mobbed up. Then Warren turned to me. 'I'll take this side of the valley,' he said. 'You take that side. Okay?'

I nodded.

'You know what you're doing?' he said.

'Yep, no worries,' I said, as much to reassure myself as him. This was my first ever solo job as a newbie musterer, and I was flooded with nervous excitement as Warren headed off to his side of the valley. In the absence of a dog, I wove back and forth, hooting and hollering the way I'd seen the others do. Then, when my throat got too sore to carry on, I dropped the reins and clapped my hands instead. This worked, plus Sergeant approved, as he could put his head down far enough to nick a bit of tussock.

Soon, the terrain began to widen and my mob spread out, and Sergeant and I had to pick up the pace, trotting left and right to draw the cattle back together. And finally, after a few hours, I dropped down into the flats. I could see Warren's mob filing forward, and in the distance Ynyr and Andrew had appeared to help out.

There are a number of river crossings on the Greenstone muster.

The craggy old door to Passburn Hut, deep in the Greenstone Valley.

In that moment, the big picture of the muster came into focus. The stretching-out of cattle reflecting the shape of the reaching-out tributaries and the main thread of the river. Horses and riders in the flow, and the day arching above it all. It was truly breathtaking.

*

Soon, we were pushing back to the trees we had burst out of on the first day, slowly mustering our way homeward. Assigned individual cuts of cattle, we spent hours quietly pushing our cows forward, the trees acting as a helpful barrier. It wasn't always easy. The cattle were getting tired, and the calves got mis-mothered, causing little breakouts. Whenever the track opened up into grassy clearings, cows would take the opportunity to make a run for it, often heading for the enclosing dark of the trees.

I stuck with Ynyr, who told me, 'You have to look ahead and think what they're going to do before they do it.' Some of the only words he shared with me as we rode together in easy silence – until, that is, he let loose an explosion of expletives.

We'd been trying to coerce the cattle up a steep and muddy slipped-out track, but they just wouldn't budge and we couldn't understand why. Then, suddenly, the cattle started turning on us, and all hell broke loose. Swearing was necessary. It was late in the day, we were tired and cold, and the weather was unrelenting. A slow, annoying drizzle had made an appearance, and as Ynyr swivelled in his saddle to face me I could see a long drip forming on the tip of his nose. 'What the heck?' he said. 'Why are they doing that?'

Before I could respond, the answer presented itself: two trampers popped up at the top of the track like meerkats. 'Hello!' they shouted, completely unaware of the chaos they had set in motion. I cringed, knowing they would not be receiving quite such a cheery greeting from Ynyr. 'Bugger,' he cursed, before shouting at them to move well off the track. They did as they were told – but not before the cattle had scattered every which way. We spent the next few hours trying to get them back together.

When we at long last made it back to the hut, we were exhausted, hungry and bedraggled. The fire was going, the billy was hot, and the rest of the team, sitting there on wobbly chairs and bunk beds, could not have been a more welcome sight.

'We thought you'd got lost,' said Andrew.

'Or that you'd buggered off home!' teased Grimmer.

'Bloody cows,' was all Ynyr muttered in response. 'And bloody trampers.'

'Here, have a beer,' Hayden said, passing one my way. 'The whiskey's run out, and we won't be carrying these out.'

The mood that evening was noticeably more melancholy than the previous ones. It was our last night out here, and there was no escaping the knowledge that the muster was finally drawing to a close.

*

Fittingly, mist had settled in the valley for the muster's last morning. It was sombre, wet and cold, and as we closed the door on Passburn Hut I know I wasn't the only one wondering when that door would be opened again.

We each took a cut of cows, and I found myself alone once more behind a mob of forward-moving animals. Into the trees we went, high silver beech on either side, and the going was slow and steady. Some hours later, the track opened up again, and mirror-like Lake Rere came into view. We stopped for a subdued lunch, and I asked Stu what it meant for the valley to have the cattle taken off it.

'Well, it will just go back to grass,' he said. 'The grass will layer up and get long. A bit more matagouri will come in.' Then he paused, lost in thought for a moment. 'The old-timers used to take the calves up to the flats. They would cut a tree down across the track, and leave the calves there to wean them. And when they had to bring the calves out, they would go and cut the tree back to get them out. All done on horses.'

'What would they think of this, eh?' said Warren. 'Taking the cattle off for the last time.'

Stu shook his head. 'There will never be cattle out here again. Seems unbelievable, really. But I fully understand it. I do think that intensive farming in other parts is a big problem. They are putting some real pressure on the land. And it's people too. People are a real problem.'

There wasn't much talk as we mounted up and made the final push home. And, as we came out of the trees and found Lake Wakatipu waiting for us, the rain really set in. We were all riding together again, as we had been at this exact point on the first day. We'd come full circle, only we'd picked up a mob of cattle on the way. The horses were pulling a little, impatient to get back to their home paddock. Their wants were simple: saddle off, food, a drink, and a roll on the ground to itch the itches. And our wants at that stage in the day were actually pretty similar. Finally, the cattle were put into a big holding paddock, and the mob spread out, lowering their heads to graze. And that was it. The last muster was done. There was no fanfare, but we all felt it. We let the rain fall on our heads, and none of us said much at all. What was there to say?

Then Stu mumbled something about feeding out and handed his horse's reins over to me, and off he went. Maybe to be on his own for a moment, or more likely just to look after his cows. Stu was the definition of pragmatic. The rest of us followed suit, dismounting and tending to our chores. Time passes, things change, but on a station there is always another river to cross and wet cattle and horses to look after. So the day goes on and station life continues, and as one chapter closes, another opens.

2
HIGH-COUNTRY HORSES

Muller Station, Awatere Valley, Marlborough

To get to Muller Station, the only way is up, up, up. Before starting the climb, I paused in the little town of Seddon, and asked a local about the road I was set to drive. In response, they offered me a question: 'You're not scared of heights, are you?'

This is the domain of mountains – big, big country charged with an elemental energy. Stretching out from coastal Seddon to Hanmer, the Awatere Valley is underscored by the inland Kaikōura mountains in the south, and to the west and north by the Black Birch Range. The highest point is Tapuae-o-Uenuku which, at 2885 metres above sea level, is the highest summit in New Zealand outside of Kā Tiritiri o te Moana/the Southern Alps, a loud

The Muller Station crew: from left Steve Satterthwaite, Mary Satterthwaite, Kelsey Green and Alice Satterthwaite.

A young pup in training at Muller Station.

shout of a mountain dividing the Awatere River from the Clarence Valley.

The road up was indeed formidable. My car felt very small, and as the tarseal gave way to gravel I gripped the steering wheel tighter, reminding myself that at least it was better than it once was. Where I now drove across bridges, there had once simply been fords, and the road had been vulnerable to flash flooding and shifting shingle. Awatere means 'fast-flowing river', and the turquoise braided waters could become an engorged torrent; the river had, at times, ruled the valley.

Making my upwards journey even more daunting was the fact that I was pretty much following a faultline the whole way. The Awatere Fault runs a sharp line along the north-west margin of the valley, and over the years it has had a say in the geology of the area, with the scars of the Kaikōura earthquake still running deep in the land.

It's this mountainous terrain that makes the valley a stronghold for horse. And Muller sits at its very top end, up where the Kennet River joins the Awatere – almost as far as you could fling yourself on the southern side. The station first gained a foothold in 1851, but its current owner, Steve Satterthwaite, took up the reins from his father, Clive, who bought it in 1965. Nowadays, Muller comprises 38,800 hectares, making it one of the largest stations in the country.

The day dimmed as I got higher, and a drifting fog reached for me with long fingers. As night fell, my imagination summoned beasts from boulders and dragons from shadowy corners. I glimpsed the occasional glow of a farmhouse, but any sign of life was far and few. In the creeping dark, I finally rattled over First Creek bridge, and my headlights picked out the sign: Muller Station.

I made my way down a long drive, crawling past horses in paddocks alongside, and drawn to the welcoming beacon of the homestead's porch light. Mary, Steve's wife, had waited up; she came out to greet me, then made me a cup of tea.

'You can meet everyone in the morning,' she said, before showing me to my bed and bidding me goodnight. Then she disappeared.

*

When I woke the next morning, I peeked out my window, hoping to get a glimpse of the view that had eluded me the night before, but the fog had lingered. This valley was determined to remain mysterious, and for me to be patient.

High-country horses – Muller Station

I could see the lay of the land right in front of me a little clearer, though. The Muller homestead sat in the basin of a river valley, a cluster of farm buildings and sheds, with staff and tourist accommodation stretching out from it, like a tiny rural village with horses dotted over the surrounding landscape. I knew there were mountains out there, too, but they still wore their feathery cloak of cloud; all I could see was the foothills, folded into themselves with stony knees bent to the river.

The dogs were soon awake – howling, whining, rattling their chains, wanting to get on with their day of work – so I got up and followed the smell of coffee to the kitchen. This part of the house was the place to be, and here Mary held court. She was in charge of the morning hustle and bustle. Mum, horse musterer, feeder of many people and general do-it-all person, she was tall and athletic, a whirlwind of productivity. I wondered how long she had been up.

Steve was already there, a proper high-country, shorts-all-year-round man, as tall as I was short, and with a gruff manner, neatly trimmed beard and quick smile. Soon, his and Mary's daughter Alice, who'd taken over the stock manager job a year or so earlier, sidled in and nodded briskly in my direction before reaching for the coffee. Then came

The big country of Muller Station with a fresh dusting of snow.

The Muller Station homestead is nestled in below rolling hills that reach into bigger heights.

Kelsey, the family's much-loved shepherd and Alice's right-hand helper, followed by Brossy, who was training at the station under the Growing Future Farmers scheme.

Everyone here moved at a clip, so I followed suit, shovelling down my breakfast. We had a big couple of days ahead, I was informed, mustering cattle and calves from various parts of the station. At Muller, things were busy year-round – there was never much of an ease-up – but autumn was particularly full on. Snow was imminent, so the cattle that had been summering in the high reaches at the head of the Acheron needed to be brought homeward for weaning, just like we'd done at Greenstone. In another month, there'd be a big three-day muster to gather cattle from the highest points of the station, which rose to a cloud-skimming 2000 metres above sea level, but in the meantime we were going to get started bringing in the cattle that were closer to home.

'Do you use horses for most things?' I asked Steve, taking one last hurried sip of coffee before following him out the door.

'Yeah, we do,' he replied. 'Horses just make sense in this kind of country.'

'And it's way more fun on horses,' Kelsey chipped in.

Alice didn't join in on our chat; she was focused on the day ahead. She handed me a halter, then pointed me in the direction of a solid-looking chestnut horse with a long, tangled mane. 'That's Summer,' she said. 'My brother Ben's horse. He's working up north in the East Coast right now, so you can ride her.'

Summer was a St James horse, meaning she was bred in the not-so-far-away Ada Valley, where horses had run as a herd for over 100 years. They were well regarded for their strength and agility, and many shepherds swore by them as the best mount going.

The rest of the team were already saddling up in a flurry of flicking brushes, clanking stirrups and jangling bits. In calm contrast, a dozen dogs were clumped together, patiently waiting on the concrete steps outside the tack shed, some watching with heavy heads on paws, others with eyes alert and expectant. I raced to keep up, flinging my battered old saddle (still a bit damp from the Greenstone rain) onto Summer and tightening her girth. I slipped on her bridle and at last, horse at my shoulder, I was in business.

Then, without a moment's pause, we were all loading our horses up into an open-topped four-wheel-drive truck, and Steve was explaining to me as I climbed up in to the cab and squeezed alongside Kelsey and Alice that Muller was the kind of land where shepherds, managers and landowners could really flex their mustering muscles. There was, he reckoned, a symbiosis between man, horse and dog when it came to doing stock work. 'You'll get to see it in action today,' he told me. 'And you'll see how quiet our cattle are.'

*

Twelve kilometres of bumpy road later, we arrived at the mouth of an old track that ran easily alongside the trickling creek. The truck had barely stopped before we were out of the cab and unloading the horses again. They clattered down from the truck backwards like total pros, then we mounted up and set off.

I was assigned to tag along with Kelsey and Alice, checking one side of the creek while Steve would check the other; meanwhile, Mary and Brossy were coming behind us and would catch up. Our trio of women followed the low track, and I kicked Summer on in order to make sure I didn't get left behind. Soon, we reached what I was told was the Langridge Block. One of the station's original blocks, it included some low-lying country bordering the river from

Castle Creek to the Molesworth boundary. The ruins of an old cob homestead and settlement slumbered quietly here, and nearby parts of an old horse-powered whim – a device used for hauling ore to the surface – poked through the long grass.

It was tough going here in the early days, when rabbits caused havoc with the landscape, and the Langridge Block copped it particularly badly. Aerial poisoning in the 1950s had eventually brought the rabbits under control, and the land had gradually recovered over successive years, but it was still a fragile ecosystem. To Steve's mind, this was yet another reason to use horses: it's much kinder on the environment than using a heavy vehicle. 'It's part of keeping things sustainable,' he'd told me earlier at breakfast. 'We try to walk lightly on the land as much as we can.'

From the Langridge Block, Kelsey and I headed higher, while Alice peeled off into the lower reaches to look for cattle. And, at last, I got my first glimpse of the high hills peeking through the fog and my first moment of pause that morning as Kelsey stopped and, still looking for cows of course, we soaked it all in.

'It's pretty good, eh?' said Kelsey, and I just nodded, silenced by the view.

Muller Station is situated in some stunning and dramatic country, right up the top of the Awatere Valley.

It was mid-morning, and the day was promising to be a beauty. A fine scree peppered the mountaintops, then gave way to tussock, shrubland and finally, at the lowest points, more fertile pastures. There was a truly untamed feeling to the land, and this whole area was a habitat for some very special and rare native birds: kārearea (the soaring eastern falcon), pīhoihoi (the New Zealand pipit) and tītitipounamu (the South Island rifleman).

'I love these hills, that's for sure,' Kelsey said, an unmistakably southern roll to her Rs – she hailed from a dairy-farming family and called the flatter lands of Cheviot, in North Canterbury, home.

As the day warmed up, we peeled off our jackets, and agreed there were no cows as far as we could see. Heading back down the way we had come, we spotted Alice below us, zipping along on Bug, her little Appaloosa cross. Bug would have won the Fastest Walk Award any day, and Kelsey and I had to chuck in a canter to catch up with the duo. Alice did everything fast, and rode with complete ease and confidence, her eyes firmly set on what lay in front of her. 'I've been chasing cattle since I was a kid,' she told me. After she'd left school, she spent time working as a shepherd in Australia, which had given her an appreciation for smaller, sportier horses – and when she got back afterwards and saw her more traditional, heavy-set horse standing in the paddock, she thought it looked like a ploddy draught horse. 'I just wasn't keen on that anymore,' she said. 'So I got Bug as a yearling, and she's been great.'

'Do you break them in yourself?' I asked.

'Yeah,' she said, keeping her focus on her work. 'I was always taught that if I wanted a good horse, I needed to make it that way myself.'

*

Having found the valley clear, we met back up with the others in time to see Brossy sauntering down the road on his quiet mount Shylo behind a big Charolais bull. Alice took over the bull, and it proceeded to give her a bit of grief, turning on her rather than moving in the direction it should have. But Bug showed her worth, not taking a second of the bull's bad attitude. Ears back meant business, and the speedy little mare had it sorted out and heading homewards in no time.

I settled in alongside Steve, who was riding Savvy, another St James horse. 'I fell in love with this high country when I first came out to look at Muller with my father,'

Steve told me as we walked along behind the mob. 'He'd come up here on a horse trek once and liked the land, and so when it was on the market his interest must have been sparked.' That was in 1965, and Steve was 13. 'Something fired up my brain that day,' he told me. 'I knew this was where I wanted to be.'

When they bought Muller, his parents were farming in Culverden. Compared with the flat plains of North Canterbury, the mountainous largeness of the Awatere Valley was so different and exciting for young Steve that he couldn't get enough of it. But his family only visited the station a few times a year, having promoted Muller's head shepherd Barry Bensemen to manage it, so when he finished school in 1971 Steve came straight up the valley to follow his dream of working the hills. 'And I got no special treatment by being the son of the owner,' he said with a grunt and a wry grin. 'In fact, it was probably the opposite. But I got to work learning all I could. I was a "cowboy", and back then that meant I was the young one who had a lot to learn.'

Barry was a 'hard man', Steve said, who came from a family with a long association with the high country. 'He taught me that a strong work ethic was necessary, and that stubbornness could be useful.'

Autumn is the time for calves to be weaned from their mothers.

Muller Station utilises an open-top truck to transport horses to the starting point of a day muster.

Our horses' hoofs were crunching on the loose gravel, and as the landscape in front of us opened up, the cattle did too, spreading out and trying to disappear back into the hills, while the heading dogs kept them in line and we continued pushing them homewards.

'Barry was quite something,' Steve said. 'Nothing could beat him.'

Alice had been making her way towards us from a valley gut she had ducked into, and after mingling her mob with ours, she too settled in alongside us.

'I remember back in the early seventies, we had Hereford cattle,' Steve carried on. 'You know, the ones with the big horns? And there was this one bull that, despite the fact the head shepherd and I both had good horses, always beat us.'

'That renegade bull managed to make its way almost up to Molesworth, and so Barry said, "Right, we'll go and get the bastard." He fired up the four-wheel-drive grader, and fashioned old hessian ropes into a great big lariat, then got the head shepherd to sit on the front axle of the grader, trying to lasso that bull.

'It gave a good fight, but we eventually got him. And we towed him home behind us, bellowing all the way.' Steve paused and looked me dead in the eye. 'Tenacity, that's what Barry had.'

And Barry, it turned out, was just one of many memorably tenacious horse riders to grace Muller's hills. Perhaps the most memorable (so far, at least) was John Shirtcliff, or 'Shirty' as he was known. Shirty owned Muller for almost 60 years, having taken it on at age 25 in 1896. He was a well-known character with a curmudgeonly personality, and a few peculiar turns of phrase: 'I say see boy', 'like see' and 'flaming hell' apparently punctuated many of his sentences. It's said he would tell people he had 30 staff – 'Ten coming, ten working and ten fired' – and he would frequently only hire a shepherd if the man could beat him in a fight. As well as the many stories about Shirty, there's even a poem written by the hawker and poet Harry Gallagher, who was known to haunt the Marlborough high country. It goes like this:

The Laird of the Muller
There's hustle and bustle when he starts a muster.
'Where are those dogs? Flaming hell!'
He takes to the stirrup with plenty of syrup,
And plenty of ginger as well!
The boys say that Shirty can ride as at thirty

And do any mustering spell.
On Muller and Langridge, his classical language
Imparts a superior tone.
He's really worth hearing, whenever there's shearing.
His diction is very high-flown.
They don't serve it hotter than Shirty the Squatter,
For he's got a style of his own.

*

In her role as stock manager, Alice was learning the ropes from her dad – and she was cut from much the same cloth as him. Working on the land was the only thing she had ever wanted to do. 'The guts of it is that I genuinely love the work,' she told me. 'The love and the passion comes from the horses, the dogs and the environment.'

As a kid, she would get out on the station with her mum and dad as often as she could. At first on the front of Mary's saddle, then on her pony, and later on her first horse, a palomino who she said was 'a total bastard'. 'He absolutely grilled me. But I think that taught me how to be gutsy.' Alice had just the sort of can-do, will-do attitude and hard-working ethic that will get you far in the high country.

The dry autumn day had really heated up, and as we carried on working the mob forward at a steady pace, Steve gave me a lesson in mustering. 'You can't bulldoze cattle,' he said. 'You have to teach them. The key is to get them stringing. With livestock, if the lead goes, the tail will follow. If you push too much, you can have a real shit fight.'

While he admitted that some days do still turn to custard, he reckoned it was usually because of something a human had done wrong. 'I had one muster years ago where the cattle were all mis-mothered from the get-go,' he said. 'We started at 6am and we didn't get to the top hut until 9pm. It was just the most horrendous battle. And that was our fault, not the cattle's.'

In contrast, he told me, there was another muster where the cattle had gone so quietly that the deer standing up on the face of the hill weren't even spooked as they passed through. 'Days like that,' Steve said, 'it's the best thing in the world.'

*

Back at the homestead, I found Mary already in the kitchen, whirling about making tea and food and doing a million

Alice is a young horsewoman with a real passion for her work.

Muller Station has varying country from rolling hills to craggy, high ranges.

other things. She was in such constant motion, and things were so constantly on the go here, that I was having trouble imagining how anyone who lived at Muller could ever feel lonely or isolated, even all the way out in the back of beyond. When I mentioned this to Mary, she said, 'I think sometimes we get more visitors than people would in town!' She knew her life here in this valley was special, and she liked to share it with others, she explained. 'The hill country gets into your bones. It settles in, and once it's there it never leaves.'

Mary's father and his father before him had been horsemen, too, and she'd learnt to ride on her dad's horse Goldie. 'We never had a pony,' she explained. 'We were thrown in the deep end with horses straight away. And Goldie was actually quite awful! But we all loved him anyway.' And, here in the Awatere Valley, her kids were similarly plonked on the front of saddles or on their ponies, and learnt life in the hills. Right from the get-go, 'the kids were a part of everything,' Mary said. 'It was the best place to raise them.'

In this horsey family, motorbikes remained firmly at second rung. They were used when needed, but horses always took precedence. 'There is a real mindfulness and quietness about taking a horse, or going on foot,' Mary said. 'It's about having a rapport with your horse and your dogs.

They are your companions, and it feels good to get a job done with them.'

As I helped her put the food on the table, the rest of the crew filed in, hungry eyes falling on the hot cheese scones with butter and bread rolls with fillings galore.

'Plus, if we used bikes all the time,' she added, 'there just wouldn't be the same comradeship with the team.'

'Yeah, we go out on the horses together and it gives us a connection,' corroborated Steve, who already had a bread roll in his hand.

'And when you get a young horse and you see it getting more confident and improving, that's a really good feeling,' added Alice. 'You get satisfaction out of that.'

'You can't get that out of a motorbike now, can you?' said Mary, tucking a stray strand of blonde hair behind her ear and grinning. Then, as she sat down for the first time that day, her husband leant towards me and muttered cheekily, 'Motorbikes are a bit of a North Island thing, I reckon.'

*

Not so much of a North Island thing, it turned out, that the Muller team was going to rule them out completely. After lunch, I found myself on a side-by-side quad bike, squeezed in beside Kelsey with Alice at the wheel, clinging on for dear life as we roared through the flatter lowland paddocks.

The two young women worked well together, so I contributed in the capacity of Official Gate-opener. Mustering the cattle on the bike was noisy and rough and nowhere near as fun as doing it on a horse. But, as Alice said, 'When you are under the pump, the bike is useful. That's when we use it.' It got the job done.

Then, in an about turn, we piled into the ute and hooned off down the road to muster another block on foot. As always, Alice set a cracker pace, tackling the first steep ascent like a mountain goat, her dogs right at her heels. We pushed upwards through borage stalks that had gone to seed, dried-up lupins, hawthorn and matagouri, and I started noticing the alpine plants that resided here – hawkweed, sheep's sorrel, St John's wort and sweet briar. Everything about this high country was extreme, including the incline. I cast my eyes about, trying to take it all in, but it was too immense: nothing but hills, mountains, rivers and sky, as far as the eye could see. Mostly, though, I was determined not to slow Alice down, even if it killed me, so I dug deep and breathed in ragged lungfuls of cold air and refused to stop.

Alice scans the hills for cattle hiding amongst the folds.

The dogs at Muller Station are an essential part of the working team.

We stopped at the highest point of the ridge, scanning for cattle. There were broad saddles on either side of us, and Growler, a keen huntaway, was given the command. He set off, pushing a few stragglers down towards the bottom of the valley and barking up another little mob. It was a long way down, but he was relentless, tirelessly pushing the cattle to where they needed to be. He worked like there was nothing he would rather be doing. When Alice gave him the command to return, he did so right away and was rewarded with a single pat and some praise. Gold to Growler. You could see he was chuffed with himself.

Then Alice put her other dog, Storm, to a trio of stragglers on the other side of us, and with them pushing down nicely we also descended the ridge and started up the next, checking to see if we could spot any more cattle. There were a few in the bottom groove of the ridges, and they too got flurried out by the dogs' barks. We scrambled down, zigzagging and long-stepping it all the way to the bottom.

'I'm going to check the valley,' Alice said, then she was off, so I started heading towards where we'd come from. It was invigorating, both to be here enveloped by the craggy hills and to know I'd survived what I'm sure had been a kind of test. Everything was glowing in the late-afternoon light, and on my own for the first time since early that morning I fell into a tired but contented daydream. I was only startled back to attention when little Bean, one of the Muller fox terriers, scurried up behind me. She must have decided I was the easier option to follow, and she wove along happily at my heels.

When I finally got down to the base of the gully, there was Kelsey. She had taken the lower angle of the ridge, pushing her mob through into the lush autumn grass. Then Alice arrived, and the three of us flopped down on the grass, chatting while the dogs tussled and played around us. The sun was low, spirits were high, and there was a welcome chill in the air as more cows trundled in from our right, pushed along by Steve and young Brossy.

It was a gentle way, this mustering on foot, but my muscles were telling me there was a limit to how much of it I'd be up to. Give me a horse any day. A companion, a partner in crime and a mane to hold onto.

*

The next morning, we were up and at it again, and before I'd had time to lament my aching muscles, I was already riding at speed down the outstretched arm of the station's airstrip.

Alice was just ahead of me, riding a sporty little chestnut mare called Hinemoa, and right beside her was Kelsey on Autumn. We cantered fast along the vast paddock, and I grabbed Summer's dreadlocked mane and leant forward into the cool air, keeping up with the two riders in front of me. They pushed each other on, grinning wildly, while the drumming of hoofs filled our ears and rose up to the hills and sky high above us. There had been snow overnight on the top peaks, and the change in season was tangible. From the airstrip, the country opened above us in a wide arc, cresting at the highest points.

Summer had a rolling gallop, the sort where the up beat of her movement felt paused, hanging in time. She stretched out with absolute joy, and when we finally slowed to a walk again I was red-cheeked and windswept.

We were working, sure, but every one of us – horse, woman and dog alike – was also having a heck of a lot of fun while we were at it.

*

Later that evening, my last at Muller, I sat down with Steve for a whiskey by the fire. Mary was still pottering around

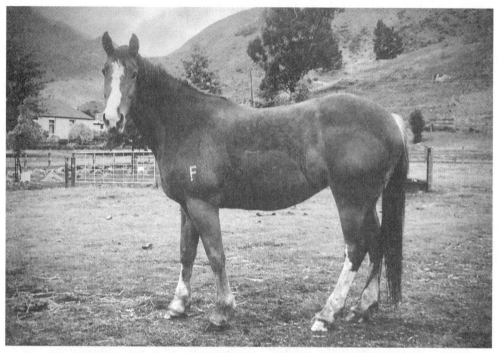

Summer is a St James bred horse well suited to the hilly terrain of Muller Station.

Everything has its place at the well-run Muller Station.

somewhere, but for the moment I relished the relative calm after my high-spirited, fast-paced, action-packed couple of days. In barely 72 hours, the team here had literally run me through the full gamut of mustering possibilities – from horseback to motorbike to foot and back to horseback again – and, in doing so, had imparted through hard-earnt experience a small sense of just why they loved using horses in these hills so much. Our equine companions were physically well suited to the terrain, but there was an emotional component there too. A spiritual one. Using horses in your work was fun, pure and simple. An expression of utter freedom and unencumbered momentum.

I wasn't the only one in a reflective mood. Swirling his whiskey gently in his glass, Steve got talking about his early days at the station. 'When I first came in to work here, life was so different to how it is now,' he said. 'Bosses were bosses, and they didn't have to be right. They just had to be the boss.'

Mary appeared to quietly put another log on the fire and top up our whiskey glasses with a wink.

From his start as an 18-year-old shepherd, Steve had worked hard and learnt fast. He'd become head shepherd in 1978, then took over from his father in 1980. 'I was lucky, in a way, not to have come from a traditional high-country background,' he said. 'It meant that I could bring some new ideas to the way this place could be farmed. And I encourage my kids to look at things their own way as well. It's like with a horse: you can take it to water, but you can't make it drink. And that's all we can do with our kids.'

As if on cue, Alice popped her head in to say goodnight.

'Thanks for putting up with me, Alice,' I said, and she just laughed.

'Ah, you did all right,' she replied with a rare but genuine smile.

Sometimes, the best way to get to know a person is to work beside them, and that had definitely been the case with Alice. She was the next generation on this land, working hard to learn while also finding her own way, just as her grandfather and father had done before her. Alice had a fierce, determined quality about her, and I knew she intended to make sure that horses would always call Muller home.

Before I too headed to bed, Steve imparted one final bit of mustering wisdom: 'Just make sure you don't stand in the gateway,' he said, a pointed finger waggling, eyes narrowing, a big grin on his face. 'That's important.'

3
THE HARD STUFF BEHIND

Awapiri Station, Seddon, Marlborough

The community in the Awatere Valley is tight-knit, and though my stint at Muller was brief, it lasted long enough for word to get round, and I was promptly handed down the valley to another station and another autumn muster. I'd already heard little snippets about Awapiri Station's forested Swale country, with words like 'isolated', 'wild' and 'breathtaking' used in the same breath. The valley's well-loved farrier, Earle, had told me, 'The Swale is a special place. If you get to go in there, you are a lucky person indeed.' So, when Sally Smith said I was welcome to join their annual merino-wether muster, I was punch-the-air excited.

I just had one question. 'Um, Steve,' I asked on my way out the door at Muller, 'what are wethers?' I realised I'd better fess up if I was going to be mustering them.

Steve smiled. 'Adult male sheep,' he said. 'They've got especially soft wool that can be spun into an extremely fine yarn.'

'Righto,' I said, glad to have asked, even if it was a bit embarrassing.

'They can be a tricky breed to muster,' he added. 'They always want to go up, and they have a tendency to bunch up and circle.'

I considered myself duly warned.

*

Awapiri Station is made up of 7000 hectares of tough country, stretched taut like a 25-kilometre-long rope between the Awatere and Clarence rivers. At one bulky bookend is the Chalk Range, a monolithic hulk crouching in the mythical Swale country, and over the gullies rise the heights of Black Mount, surrounded by remnant native bush and untouched beech forest.

The Smiths' house stood on the original homestead site, and was a modern build that incorporated some of the old cob. A large macrocarpa-lined hill rose up behind the house with sheds, dog kennels and a chook run nestled at its base, and the house itself was as practical as its owners. Sally and Eric Smith were no-nonsense people, fiercely self-reliant, the resilient caretakers of this wild country.

Sally's parents, Bev and Graham Black (or 'Blackie', as he was known), had started off their married life at Awapiri, and it was Graham who had established the merino stud here. Merinos are a hardy breed of sheep, well suited to the high-country tussock land of the South Island, and since their introduction in the 1840s, the New Zealand merino has become a distinct type. When Eric and Sally took over at Awapiri, they inherited its sheep and its challenging country. A big old trek was necessary to get the merino wethers out to the summer grazing way out the back of the Swale, and then another was asked for in autumn to get them home again. The autumn muster here, much like that at Greenstone, was a revered annual tradition that always followed the same pattern: a four-day horseback mission out to the Chalk Range at the south-east corner, at the top of which Sally informed me that you could see all the way

The working dogs have a rest in the sun before a sharp descent towards the Swale Hut.

The craggy folds of the Swale country.

out to glistening Te Moana-nui-a-Kiwa, the Pacific Ocean. 'That is, if you aren't scared of heights,' she added. 'It's pretty gnarly up at the top.'

For the muster, Sally lent me a swag that used to be her father's. 'I'll show you how to roll it,' she said, then gave me a much-appreciated lesson in wrapping it round everything I would need for the next few days. Her dad, she told me, had farmed Awapiri for 40 years. He'd been brought up in Fairlie, where things had been hard because money was tight. After leaving school at 16, he tried his luck as a shepherd in the Mackenzie Country, in the central South Island, then put his foot into the stirrup on Mesopotamia, Clent Hills and finally Erewhon Station. There, he'd met Bev, whose family was farming the legendary station at the time.

Awapiri was Graham and Bev's big step into having something of their own, and their first few years were a hard slog that saw Graham spending a lot of time out the back in the Swale country. He loved those hills – they were where he was happy. They were also what took him in the end. In 2011, his car missed a tight bend on the Awatere Valley Road, and he died near the river he knew so well. After that, Bev remained very much a part of the station, but she moved into a cottage just down the road, and Sally and Eric took over.

'It's an honour to use your dad's swag,' I told Sally.

'It's seen plenty of adventures, that old thing,' she replied.

As I picked up the old canvas roll, with its soft leather straps, I could just see it being carried on horseback through mountain passes and over rivers wide, adding a bit of warmth and comfort on every back-country adventure.

*

Once we were all packed up, everyone piled into Eric's truck and set off for Flynn's Hut. To get there we headed along a track on the neighbour's property, which ran through rolling green hills. 'A bit different from our land,' Sally said. 'Awapiri is the hard stuff behind.' Then she grinned. 'Our kind of country.'

Awapiri was not for the faint-hearted, ruled by jagged mountains and vast swathes of mānuka, kānuka and mountain beech. Flynn's Hut, when we reached it, was nestled into a tight bend in the Medway River, and flanked by the toe of a steep scrub-covered hill in front and greener hills at the back. It had been built by Sally's parents in 1973, and while still being characterful boasted a more open design than the huts I'd stayed in at Greenstone. It looked solid too, with flat iron cladding over pine framing – and it even sported a tongue-and-groove floor, a huge wood-powered Shacklock stove and a deck.

The horses were already there, having been ridden out earlier, and they paused their grazing to eye us suspiciously, knowing work was on the horizon. Here, I was introduced to my ride for the next four days: Copper, a wide-shouldered chestnut quarter horse who Eric reckoned was 'a bit in love with himself'. Quarter horses were first brought into New Zealand from the USA in 1969, but the breed originates from the Spanish barb horses that the Chickasaw Indians cultivated, which were then crossed with English stock horses imported to Virginia. Quarter horses were traditionally used in North America for ranch work because of the breed's amazing speed over short distances and natural ability to outsmart stock. These horses are built like bricks, with hefty shoulder muscles and handsome blocky heads, and Copper was indeed a good-looker.

Nearby was Memphis, Sally's reliable and trusted mare who was a bit of a mixed breed with perhaps a touch of draught. Black with a white blaze on her teddy-bear face,

The mist moved in. Weather changes quickly in the Kaikōura ranges.

Mist shrouded hills with a hint of sun. Awapiri Station has some dramatic views.

'She can be a little prone to falling asleep,' Sally told me, as Memphis softly nuzzled her hand. Sally had spent her life around horses, as her dad used them to muster all the time. 'He would head out with two horses, and when one got tired he would ride the other,' she said. Like everyone else I had met up the valley, Sally had learnt to ride by just getting biffed on. 'We had a pony and three of us kids had to share it, so as soon as we could, we would figure out getting on the bigger hacks. They were much easier to handle,' she said. 'From quite a young age, we would bugger off for half the day, tearing around.'

Back in the day, horses had been a necessity on Awapiri, and for certain parts of station life they still were. 'We like using them,' Eric said, passing me Copper's bridle. 'Always have. And it does make things quicker out here.' Awapiri climbs from 350 metres above sea level to 1500, and most of the station's tracks are not suited to vehicles. This means the fuel bill on Awapiri is tiny, which suited Sally and Eric in more ways than one, as keeping things low-cost was a priority. They weren't big-scale farmers, and preferred to do most of the work themselves.

For his part, Eric had come to horses a bit later than Sally. It was at Erewhon that, as well as meeting his wife, he'd really become a horseman. 'I was there when Sally's uncle had it,' he told me, 'and there was one day when he wanted me to go over the river to do something. I hadn't been there for long, and he said, "Can you ride a horse?" I told him I'd ridden my sister's pony when I was a kid, and he just said, "Well, there's no time like the present. Go and catch that horse, and we'll chuck the saddle on it."'

Sally, overhearing this, laughed; she knew straight away what Eric was talking about. He just grinned his sideways smile and continued: 'I got on and did a few laps round the yard before he decided I would be fine, and off I was sent over the Clyde River.'

'And that's not a small river!' Sally added.

Eric shook his head. 'No, it was quite a big river to cross for a newbie.'

But Eric survived the day, and those that followed. He and Sally went on to work at Blairich, Fairlie, Grays Hills and Shenley stations, before eventually ending up in charge of Arrowsmith Station, on the shores of Lake Heron in the central South Island. They spent a decade running this station, which became a snowy, icy fortress in the winter. Water froze in taps, dogs howled at 3am from the cold, and winter chores were a daily endurance. And for two of those

GREENSTONE

A moment of rest in the sun at Greenstone Station after the last cow is pushed through the gate.

Bush Creek Hut is at the far boundary of Greenstone Station, set back from the valley.

GREENSTONE

A string of cattle are pushed up the valley towards the homebound track.

Planning the next move as the musterers wait for the last mob to be pushed past.

GREENSTONE

An open fire is a welcome comfort in a back country hut as well as being the place to cook meals.

The Rats Nest Hut is at the top boundary of Greenstone Station with Milford not far beyond.

MULLER

The Muller Station homestead has a pretty enviable view over the Kaikōura range.

MULLER

Kelsey is a skilled shepherd and she's pretty handy on a horse.

Steve checks for cows hiding away in the crevice of a hill.

AWAPIRI

Sally Smith is very much in command of the Awapiri muster and her knowledge of the land has been hard-earned.

Merino sheep are a hard sheep to muster as they tend to circle rather than move forward.

AWAPIRI

The terrain of Awapiri Station is not a place for quad bikes, but it is passable with a horse.

AWAPIRI

The horses of Awapiri are a working tool but also well-loved members of the station team.

There is plenty of walking involved in a muster. Down hills are a time to hop off and give the horses a break.

TIROROA

Tim Mansell still plays an active role on Tiroroa.

Tim looking for stray sheep in a gully.

RUANUI

Mataroa Hall is a classic old building on the road out to Ruanui Station.

RUANUI

Jo doing the hard yards gathering up cows on Ruanui Station.

Rylee doesn't miss a chance to help out on Ruanui Station.

RUANUI

The manager of Ruanui Station, Jeremy, has spent a life around horses.

Father and son move the cattle back to fresh pasture.

RUANUI

A day of pushing cattle into the yards for weaning is made easier when there's blue skies behind the clouds.

Jeremy uses his horse to separate calves from their mothers.

SMEDLEY

The rolling hills of Smedley Station with gullies of mature native trees.

A group of senior cadets at Smedley Station work on training their young horses.

SMEDLEY

The top reaches of Smedley Station still retain areas of bush.

The horse barn at Smedley Station is a place for cadets to get their horses ready for a day of work.

SMEDLEY

Smedley Station cadets work together during the springtime lambing beat.

Past Smedley cadets make their mark on the barn walls.

The hard stuff behind – Awapiri Station

ten years, Eric and Sally ran a massive 14,500 stock units on their own, just the pair of them. That's the way they liked to do things: off the sweat of their own two backs. 'We loved it,' Sally said. 'The challenge of it all.'

It certainly prepared them for taking over their own station. And, having used horses throughout their working lives, they of course kept the home paddock at Awapiri as the horse paddock when they arrived. 'It's what we know and enjoy, I guess,' Sally said. 'It wouldn't be the same here without them.'

*

It had been a while since I'd had to tack-up with a western saddle, and I had a red-faced struggle getting the extremely heavy bit of kit on to Copper's back. The riggings of western saddles are different from English saddles; they're designed for riders who spend many hours mustering and have a deeper seat, longer stirrups, and a horn where a rope can be hung. These saddles are held in place by cinches that, rather than a buckle, are done up with a series of loops that, to the untrained, can be as complicated as origami. In the end, I had to get Eric to help me out. The saddle had been his for about 30 years, and it was a beautiful worn acorn brown. 'It helped get a truck out of the mud once,' he told me. 'We tied a rope to the pommel and got the horse to pull.' While he redid the girth, I tried to memorise the technique.

Copper performed a little dance and pushed past Sally and Memphis as I mounted awkwardly, negotiating both the pommel and the long hill stick in my hand. I hurriedly tried to collect myself and my horse. As always, there was no time on a muster for hold-ups and no special treatment would be given if you weren't up for the task. Fortunately, Copper turned out to be a chugger, in the true style of a quarter horse, and once I settled into the depths of the western saddle we quickly became a working pair.

Our merry band rode out and up, waving goodbye to Eric; he would catch up with us later at Swale Hut. There was always so much preparation to do before a muster, and it was a relief to be finally moving forwards. We started along the old Awapiri Pack Track, which had first been recorded on a map in 1917. It was carved out of the rock, and the going was predominantly up. With every kilometre we gained, the drama also heightened: we were in the domain of steep mountain slopes and craggy bluffs. The horses' hoofs clattered on the greywacke track, and rocks would

A string of horses walk single file through ancient Mānuka.

From mist to slanting sunlight.

occasionally dislodge and tumble through deep gullies below, crashing down until we could no longer see them. Here, at the north-east end of the Inland Kaikōura Range, the active uplift of the range had formed steep mountain ridges and knife-edge valleys.

Single file, the horses put their backs into their work. At the front was Sally on Memphis, followed by our packhorse, Bonnie, whose lead rein had been unclipped due to the narrow trail. Packhorses also played a big part in New Zealand's history. Their precursor was the journey horse – hardy horses, strong and compact, that carried riders and their provisions from coastal areas to the country's wild interior. Later packhorses carried all sorts of things on their backs, including the pianos of British colonists, and were a necessary means of carrying gear in to the high country on long musters. These pioneering horses were often tough characters, nuggety and strong, but their presence started to dwindle after the Second World War. Returning farmers who had experienced the heft of American bulldozers while serving overseas started to use the burly machines to open up wider tracks on their land. Places that were once only accessible for horses became the domain of tractors and jeeps, which slowly took over the work of the mighty packhorse. Here on Awapiri, though, a packhorse was still the best, and only, way to go.

Behind Sally, Memphis and Bonnie came Sally's daughter-in-law, Bridget, who was joining the muster for the second year in a row. Bridget was on Raisin, a gorgeous chestnut pony with a golden mane and tail that shone like freshly cut hay. Locally bred, from just up the road at Upcot Station, Raisin was a bit of a happy accident: the result of one of Upcot's mares jumping in with the Upcot stallion. She'd been Bridget's first breaking-in project, and was still learning the ropes. It was the pair's first muster together. 'She'll have to figure her feet out on the steep bits,' Bridget had said to me earlier. 'I just hope I stay on!' Now, little Raisin was negotiating the tough terrain with increasing confidence. At one point, she had to contend with a scramble over an open scree face, and Bridget was left white-knuckled but proud of her young pony. 'Phew!' she called behind her. 'I think she's officially found her feet!'

Behind Bridget and Raisin came Andy, who was on Oak, a big and proud-looking gelding who started life at Molesworth Station at the top of the valley, so therefore took Awapiri's huge hills in his stride. Andy lived at neighbouring station Glen-Lee, his family farm, and he always lent a hand on Awapiri's autumn wether muster. He knew these hills just as well as Eric and Sally, and had the big hands and solid calf muscles of a proper high-country man; when he'd turned up with his team of dogs at his heels the day before and introduced himself, his handshake had been wince-inducing strong.

Copper and I held up the tail end, and after many hours of riding (and, on my part, concentrating on staying on) we reached the highest point: 1100 metres above sea level. As we stood there on the Pack Track Saddle, so named by Sally's dad, we watched a kārearea soar on the updraughts, quietly taking it all in. The horses and dogs even seemed to be awed by the view. The saddle sat between Black Mount and Mount Malvern, and provided a 360-degree spectacle of absolute wildness. We were just tiny dots in the midst of an enormous landscape. The inaccessibility of this mountainous terrain meant the whole area was basically untouched, with impressive tracts of diverse native scrubland and forests of mountain beech, mountain tōtara and kānuka. And, to my far left, I could just make out the tip of Tapuae-o-Uenuku winking at me. The maunga's name poetically means 'footprint of the rainbow', and it can also be seen from Kāpiti Coast, all the way across the sea and not far from

Sally Smith is a true high-country woman who loves her work.

Up, up, up and then some more. There's not much flat country at the back of Awapiri Station.

my home in the Manawatū. Captain Cook nicknamed it The Watcher because he felt like it could see his every move, and after Sir Edmund Hillary climbed it very early on in his career he pronounced that he'd 'climbed a decent mountain at last'. From up here on the saddle, I felt similarly.

'This view never gets old,' Andy said, breaking the silence of our shared reverie.

I looked over towards where I knew we needed to be heading, but all I could see was a steep, rocky spur falling abruptly to the valley floor. 'Where does the track go from here?' I asked. Everyone laughed. 'Come here and I'll show you,' said Andy. He led me to the edge and pointed. 'See that zigzag track there?'

'Yes …' I said hesitantly. I could see the track … but I couldn't see how we were going to travel down it. It was shaped like a lightning bolt, as skinny as a sheep path, and the pointy end of each carved-out V where the track pivoted was pin-tip tight.

'Don't worry,' Andy said, registering my concerned expression. 'The horses know this track well.'

Thankfully, he was right. Sally headed down the impossible track first, dismounting and throwing the reins over Memphis's head in order to lead her down, hill stick at the ready, and Bonnie came right behind them both. A hill stick is longer than a walking stick – roughly the same height as the person wielding it – and traditionally made out of sturdy native wood. Shepherds always keep theirs close, and their sticks become so well used they end up shaped to their handholds. You use a hill stick as a third leg, leaning into the hill and letting it steady you when your feet start slipping beneath you on scree or loose rocks. 'The best way is to let yourself slide and see how it can save you,' Sally told me.

Like many things in the high country, there is an art to using a hill stick, and I had still to learn it. Clumsily, I followed the train of musterers and horses down the terrifying zigzag, while everyone else – horses included – navigated the tight turns with astounding surety. Bonnie even put in a trot near the bottom, overtaking Sally and Memphis, our swags swaying precariously on her back; then she stopped abruptly to lower her head and start eating a great mound of tussock grass. I guess she must have had her eye on it. Copper and I reached the stony valley floor a little slower, but we did make it, and while my horse enjoyed a long drink from a trickling tributary of the Swale Stream, I breathed a sigh of relief.

*

Still leading our horses, we soon came to the edge of an ancient beech forest. Here, an old lichen-covered fence-post marked the trail. 'I'll go check the sheep track, Sal,' said Andy, handing Oak's reins to Bridget. The sheep track ran parallel to the pack track we were on, and was narrower and steeper. It was where the merinos, who were scattered throughout the Chalk Range and its lowlands, would eventually begin their journey home.

Andy set off, and I followed Sally and Bridget into the trees. Mānuka reached its charred-looking arms overhead, tips touching to form an arched cave. Copper's hot breath warmed my shoulder, and all I could hear was the soft thud of hoofs on compacted earth, the occasional mimicking call of a korimako, and now and then the bone-break snap of a branch. We stopped once to cut back fallen branches – Sally carried a small saw on her saddle and made short work of it – but otherwise we continued on quietly.

Awapiri was showing me a different, less dramatic, but still remarkably remote face. This was the Swale Country I'd heard so much about, and it was everything I'd expected and more. Not many people ever passed through here;

besides Sally, Eric, their seasonal helpers and the occasional hunter, no one had much cause to. Here, in the depths of the forest, there was very little human interference.

Sally looked over her shoulder at me. 'It's peaceful, isn't it?' she said.

'It is,' I agreed. 'The rest of the world seems so far away.'

'Well, that's because it is,' Sally said. 'Thank goodness. What needs to be done next week or what I should have done yesterday doesn't seem relevant out here, eh?'

'Nope, it really doesn't,' I agreed.

*

By the time we arrived at Swale Hut, we had dropped back down to the same altitude we'd begun at. The late-autumn sun was generously hitting the hut's shining silver sides and welcoming smoke was wafting from the chimney. Eric was standing at a hitching post waving, and Andy soon came out of the hut and joined him. 'What took you so long?' Andy hollered cheekily.

Swale Hut was to be our home for the next few nights. It sat right at the isolated southern end of Awapiri, on a grassy terrace with the Swale Stream as its neighbour. In this same

Hitched up outside the Swale Hut.

A well deserved moment of pause for the whole crew.

spot, an older pole-beech framed hut had once stood, but this newer six-bunk hut had been built by the New Zealand Forest Service in 1977, and Sally and Eric had done a heap of work on it since, including adding a deck. The view from this deck was layered with hills, and the lowering sun picked out the many shades of brown, grey, copper and bronze. Mount Tapuae-o-Uenuku, still watching, even gave me another wink from far off in the distance, before I headed in to offload my swag, get some food and hit the hay.

*

It was still dark when we set off the next morning, and I was glad for it. We were mustering on foot today, which meant our hill sticks were back in hand, and the darkness thankfully hid my fumbling progress.

Eric was on pack-up duty, so the rest of us had left him at the hut and set off at a cracking pace, as usual, with Sally up front and the dogs excitedly buzzing around their masters' feet. Once the sun had come up, we stopped by the creek, and the intrepid trio each got their binoculars out, surveying Mead Hill and the beginnings of the Chalk Range up ahead. The part of the range I could see was a beast, its giant vertebrae rising from the ground. With its white colouration, the range had a moonscape appearance, and its high cliffs and faces were extreme and barren. Beneath it spread a more habitable environment, where shrubland gave way to open grassland.

'Can you see them?' Sally was asking Andy and Bridget, and there were nods all round.

With my naked eye I couldn't see anything, but apparently there were merinos scattered up in the top reaches. Plans were hatched and Sally issued commands: Bridget was sent to take one corner punctuated by a momentous hill with a scar of slippery scree, and Andy was sent to Sleepy Peak. 'And I'll go up,' said Sally, by which she meant she would be tackling the top ridge of the Chalk Range, the boundary separating Awapiri from Bluff Station. 'From up there,' she told me, 'you can see all the way to the coast, sometimes to the North Island. And on the other side is a whole series of ranges that go out to the Clarence.'

As for me? I was delegated to act as a human gate for the merinos that Bridget would push down towards me, then turn them up a scree face.

Sally set off, and I gazed up at Andy's beat. The high perpendicular cliffs didn't exactly match their name. 'It's a shit of a climb,' Andy told me, before heading off too.

The hard stuff behind – Awapiri Station

I trailed behind Bridget, until she too went her own way. Soon, my three companions were dots in the distance, and I had a little time on my own. From the basin where I stood, stretched the wide expanse of a now dried-up lake bed; Sally had told me the lake had formed here in 2020 after a big slip created a natural bowl that filled with icy mountain water. It had been full and deep and, that year, Sally had swum across its cold waters.

Before long Bridget's voice crackled over the radio: 'I've almost got them down. Are you ready?'

'Ready,' I replied, squinting into the distance at the sheep heading downwards. I could see Bridget stepping skilfully down behind them, and then the sheep burst out onto the basin's floor. I coerced them with hoots, hollers and flapping arms towards the scree face.

'That went well,' I said to Bridget when she caught up with me. Her face was flushed, a dribble of blood on her arm, a fresh tear in her shirt.

'It did, eh?' she said. 'I was surprised. It all went so good.'

Bridget lacked the bravado of some shepherds her age, and I felt a bit motherly towards her humble pride.

'It was absolutely perfect,' I told her.

The Swale country at the back reaches of Awapiri Station is a stunning landscape.

A tightly bunched up and controlled mob of merino sheep.

The hard stuff behind – Awapiri Station

*

I stuck with Bridget for the next few hours. We made our way up to the Stag Antler Saddle, and I was stunned at just how much land there was between us and the hut. Layers and layers of shrubland hugging gorges and bluffs backdropped by the aptly named Razorback Ridge. I was glad I wasn't negotiating it on my own. Having knowledge of this land comes only from being in it. Walking every hill and noting every crevice. Putting the pieces together and understanding how things connect as a grand and magnificent whole. Eric and Sally knew this terrain so intimately because they had practically become a part of it. They knew the best ways to muster the sheep because over the years they had tried all the ways. And they knew all the tricks the merinos pulled.

Everyone had been out for about six hours by this point, and the three separate mobs were heading in the same direction. Everything was coming together. I was sent up a hill and into a gully, and told to act once again as a barrier when the sheep eventually sidled around to the other side. After a scramble through the undergrowth, I found myself with an extended rest: 40 minutes huddled in a matagouri bush, above a meditatively trickling creek. This gave me time to assess my blisters and notice the finer details of the environment I was in. Here was moss, some woolly, some wiry. There was the flat, broad leaf of the Marlborough mountain daisy, and over there some white clover and a tiny native violet. The day had been driven forward at a fast pace, but for a moment it was still and peaceful in my little windbreak.

Then a voice stuttered over the radio, warning me the wethers were near, so I ignored my complaining muscles and got to my feet, then flapped my arms as necessary. The sheep politely went in the right direction, jumping the creek and filing up the hill like well-behaved schoolchildren, with Bridget, Sally and Andy filing in behind them.

It had been a successful day for the musterers: they'd gathered about 1300 wethers from impossible places. 'You're all amazing,' I told them, as we ambled back up the hill and the hut came into sight.

'All in a day's work,' said Sally. 'And a good one at that.'

*

The next day, we pushed the wethers one more step homeward, making once more for Flynn's Hut. Back through the bush of the Swale country, and back up the impossible zigzag, which was no less daunting from the bottom than it had been from the top. The day was warm and golden, with Sally and Andy on foot, and Bridget and I taking the higher ground with the string of horses. Every now and again, Bridget jumped off to send a dog out, and my role was to keep the horses behind the action.

Sally was very much in control – it really was incredible, the way she could foresee what the sheep were going to do next. She was so familiar with the breed's particular ways of heading up high ground and circling in a swarming clump, that she was like a merino clairvoyant. Whenever the hills arrowed steeply downwards, with a creek at the bottom and another sharp rise on the other side, the merinos were prone to smothering, which put them at risk of suffocating to death. Sally knew the trouble spots – anything that could go wrong already had on previous musters – and the well-oiled team was on high alert and quick to act. A cry of 'Smother!' would ring out, and everyone would drop what they were doing, racing to the bottom as fast as possible to pick sheep up and haul them back on to their feet.

As we slowly re-covered the same ground we had ridden in on, approaching Flynn's Hut, the country became

Trusty packhorse Bonnie loaded up with all our swags and gear.

The homeward bound dash with Bonnie the pack horse being relieved of her swags.

less steep and a little tamer. The wilds of the Swale were behind us, but the view from the lookout was imprinted in my mind and my heart. Soon, we saw smoke ribboning up into the blue sky: our signal that we were just about there. After leaving the tired sheep behind the first gate we'd seen in days, we made our way to the hut, where Eric was waiting and had already lit the fire. It was a weary crew that unsaddled and went through the routine of making the horses comfortable. Chores were done, wood chopped, dogs seen to, and then finally we each sank, satisfied, into a chair, a beer in one hand and stories from the day's adventures at the ready.

*

The following morning, Sally set off on foot to push the sheep up and over the shrub-covered hill towards the track home, while the others took the ute to get out in front of them. I was tasked with getting the horses up the hill, and Copper and I had a wind-whipped and clattering canter to the top with Oak, Memphis, Bonnie and Raisin thundering along on our heels. Relieved of her pack saddle and swags, which had been shipped off in the ute, Bonnie was a packhorse transformed into a creature of flight, snorting and prancing with her tail held high.

We arrived at the top of the hill sweaty and wild-eyed, and there met up with both Sally and Bridget, who helped me to calm the horses after their mad dash. In a paddock of sweet grass, the wethers were fanning out from their tight mob to enjoy it for the next few days.

All that we had left of the muster was the 12-kilometre walk home. It was downhill all the way, so we led the horses rather than ride them. The wind was starting to stir up, and dark clouds hung heavy on the horizon as we settled into our own individual rhythms. With no sheep in front of us, the pressure was off, but the closer to home we got, the more life started to pile back in. There would soon be mobile reception, Wi-Fi and all the things we had left waiting. In the distance, we could see Andy and Eric making their own winding way back in the ute, and I imagined they were feeling the same press of worries returning. At breakfast that morning, Eric had said to me, 'It's always on the last day that you start thinking about all the other things. The stresses. People don't always understand that farming can be really hard on your mental health. The picture that people paint of how we live isn't always how it actually is. There is always something to worry about, always something that wakes you up at night.'

It's true that the high-country life comes with its own unique challenges, but in my time at Awapiri I'd also learnt that Sally and Eric weren't ones to walk towards easy. Theirs was a marathon mentality: they were in for the long haul, no matter what was thrown at them. This was a hard life, but it was also an adventurous one. Only the intrepid need apply.

My time mustering down south was, for now, coming to a close. On these high-country musters at Awapiri, Muller and Greenstone I'd seen the worth of a good horse and had got to meet the brave souls who chose a slower, more deliberate way of life, one grounded in respect and hard work. Mustering on these stations was about more than getting a job done; it was about following a pull towards adventure, tradition and passion.

4
LAST ONES STANDING

Tiroroa, Ōtaki Gorge, Kāpiti

Driving to Tiroroa, the last remaining working farm in the area, was like travelling back in time. Leaving the hustle and bustle of Ōtaki's main street, with its outlet shops and Saturday-morning market, I rattled over the bridge where the wide river travels seaward, and things slowly started to feel a little less suburban. And, just at the point in the road where the country started to feel like a deep breath, where the letterbox numbers were in the thousands and where driveways were long and tree-lined, I found the Mansells' gate. Established a century ago, 400-hectare Tiroroa sat in the folds of the gorge holding tightly to its heavy handful of history. Here, there were still horses in the paddock, dogs tied to the fence, and a dusty ute parked up at the white weatherboard house.

The Last Muster

'It's a handkerchief farm now really,' Tim Mansell told me after I'd introduced myself. 'And it's a part of us Mansells, I guess, stubborn as we are!' A third-generation farmer, he was as tall as he was charismatic, with a handlebar moustache and a cowboy's swagger. 'We're the last ones standing. The only silly buggers still trying to farm out here.'

While everyone else around them had slowly and surely subdivided for lifestyle-blockers, the Mansells had remained – and so too had their horses.

Nearby, Tim's 87-year-old dad, Barry, was busy getting his horse ready. He and Tim were to spend the morning mustering, and I was tagging along. When I asked Barry how long he'd been working on horseback, the math eluded him, so he just ended up telling me 'longer than you've had hot dinners'.

Tim chuckled. 'Come on, old man,' he said, hopping on to his big grey horse with its heavy stock saddle. 'On you get.'

'I'm just the boy here now,' Barry said to me with a cheeky wink, as he hoisted himself into the saddle with seeming ease.

I eyed his little chestnut roan, Nāti. 'Was Nāti bred up the coast?' I asked. The horse had a beautifully quiet and steady nature, and his name and appearance hinted at East Coast origins – nāti horses are those bred by the Ngāti Porou iwi, and their bloodlines are old, going back to the horses first gifted to Māori chiefs back in the 1800s. These laid-back horses are a mix of breeds, with Clydesdale and thoroughbred lines.

'Yes,' confirmed Barry. 'He's originally from Ruatoria. They know how to breed them up there. You can trust an East Coast horse. They're sensible, and they have good brains. Nāti's not as flashy as the horses I used to like, but he looks after me. He's proving himself to be a great last horse.'

'Last horse, eh, Dad?' said Tim. 'There'll be a few of those sitting around in paddocks now, doing nothing. People just aren't using them to muster like they used to.'

'It's sad, isn't it?' said Barry, shaking his head.

Tim just gave him a firm pat on his shoulder. 'Right, let's get on, eh?' he said simply.

*

Barry had grown up at Tiroroa, and throughout the years horses had remained his preferred mode of on-farm

A well-stocked tack shed full of stock saddles used for mustering through to English saddles for days out hunting.

Tim heads to the hills in search of sheep.

transport. 'The bike is so much quicker when you are on easier country,' he told me as we rode along the track together. 'But when you have steep country, unless you are very well tracked, you really need the horses. It's much safer – the horse looks where it's going, so you can look down the sides and see what the dogs are up to.'

As we headed up the track together, I watched father and son. With his dogs at his heels, Tim and his horse cut a fine form. He was a bit more rough around the edges than his dad – Barry was more English gent than charming cowboy – but horses were the pair's obvious connection point. 'And Mum is a beautiful and very accomplished rider, too,' Tim informed me.

In fact, I was riding Sue's horse that morning, a sturdy grey Cricklewood Station mare called Kate. Cricklewood, way up the East Coast in Nūhaka, was also renowned for its horses, all with a heavy dose of Clydesdale in their blood.

When the track came to a fork, Barry headed down to a lower point on the farm, while Tim and I carried on up to the heights. For a moment, I watched Barry and Nāti slowly and happily make their way, with Barry swinging his riding stick in time with his horse's steps, then I turned and followed Tim. Up on the hill, we could see all the way out to the Kāpiti Coast and beyond. It was strange to think there was a substantial town just below, and Te Whanganui-a-Tara/Wellington not too far away, either. This certainly wasn't the high country I had just come from, but it was still another world – one where the hills were still farmed, horses were still held in high regard, and things were still done slowly and traditionally instead of quickly.

These hills, the Tararua Ranges, were an important taonga for the iwi, Ngāti Raukawa. Heavily fractured by faultlines and the Ōtaki River, the ranges were used by hapū as a route to and from the Wairarapa, and as a source of food, water and spiritual power. They also provided sanctuary from enemies, and there are signs that Māori lived in the area and used horses to travel from the gorge to Katihiku pā and Rangiuru pā.

Before it was cleared for farming, the area was heavily bush-clad, with tōtara, rimu and miro, and it remained untouched until 1863, when surveyors started to eye it up for European settlers. Not long after, the gold fever that was sweeping the rest of the country found its way here, and several attempts were made to explore the reaches of the fast-flowing river, but the river took a few lives without giving up any gold.

Tracks cut by Māori were extended, and roads soon took hold. Then a start was made on clearing the rough country, with much of the tree-felling done by axes, slashers and cross-cut saws. English grasses were spread by hand to create the first pastures, and from 1889 blocks of land were put up for sale by the Manawatū Railway Company. Many of these land purchases were made from afar, sight unseen, for about eight shillings an acre.

Milling operations were opened next, but at every step of the way the wildness of this place made human endeavour difficult. It was only the hardiest souls who really managed to establish a semblance of a life here during the late 1800s and early 1900s. Some of the early European settlers lived in tents and makeshift whare, and many ended up leaving their land and walking away with nothing. Ian Arcus's family settled in the gorge in 1931, just before the Mansells, and in a little history Barry wrote, Ian is quoted as saying those times were 'extraordinarily difficult, with land of the poorest quality, the weather extreme, the road often impassable and the economics of farming almost non-existent'.

Like everyone up the gorge, the Arcuses relied on horses. Their farm was big – about 10,000 acres, and 7000 of that was still dense with bush. They had two herds of mares and

Barry on his trusty and much-loved horse Nāti.

Just like his father, Tim loves nothing more than to be up in the top reaches of the farm.

foals, and a steady supply of good farm hacks, as their back country needed mustering twice yearly, and horses were the only way in. The area's Māori communities also continued to rely on horses, and the Ōtaki-Māori Racing Club was established to celebrate the hōiho tradition.

Horses had always been a part of Barry's life, and his father had been a keen horseman, too. 'Dad was never without a horse, and we have all just continued on with that.' All of the Mansell children had ponies, Barry told me. 'We had a Shetland pony who would be harnessed onto a sledge for excursions into the hills and bush.' The Sledge Track was still on the farm. Back then, Barry would catch his horse and join his younger sister and friends from neighbouring farms, galloping from one place to the next. On summer weekends, they'd take a cut lunch and ride their trusty ponies up to the top of the gorge, or over to the Mangaone. 'They were choice times,' he recalled. 'Canoeing, swimming the horses and generally loafing about.'

When he was ten, Barry was given a pony called Chaka, who had been broken in by Charlie Arcus, a good horseman who rode the pony up the gorge on his lambing beat, fording rivers and covering at least 15 kilometres a day. 'It was a great start for a young pony,' Barry said, 'and Chaka was just magic. A bloody cracker. We formed a relationship that taught me most of what I know about horses.' Chaka and Barry spent from dawn to dusk racing around the hills. 'It was a great place for me, my pony and my dog Tip to chase hares and ducks and lord knows what.' He and his brother would ride their horses to the blacksmith at the bottom of School Road to get their shoes on, and they were sometimes joined by neighbouring children doing the same. 'It could be quite the event.'

Back then, on farms like Tiroroa, horses had many uses: recreation, transport, mustering stock, and ploughing the more fertile lower paddocks. Barry remembers his family having four draught horses, which were used for pulling the machinery that mowed and carted the silage, and for ploughing, discing and brush-harrowing the paddocks. The Mansell and Arcus families would help each other out in such tasks, and Barry said the gorge community as a whole was a wonderful thing to be a part of. 'Neighbourly cooperation was paramount,' he told me wistfully. 'One of my greatest memories is haymaking where everyone pitched in to help. It sticks in my memory as something special. Those days have all gone now, but, gosh, it was something back then.'

*

Once Tim and I had fossicked out a mob of ewes, we started pushing them down to Barry, who was waiting below. While his working dogs made quick work of tightening up the little mob, Tim and I made our way down from the hills. Then, for the final few kilometres back to the yard, I rode alongside Barry, whose cheeks were flushed from the blustery southerly that had blown at our backs all morning.

'My dad, Terence, came here in 1933,' Barry told me. 'Tiroroa was much bigger than it is now.' Back then, it had been a 1150-acre sheep and cattle block, and while much of the clearance had been done on the hill country, Barry said it was still 'pretty rough around the edges'. He told me how, as a child, he'd helped with the back-breaking job of picking up large rocks from the paddocks. 'We spent a lot of days doing that kind of work,' he said. 'The flats had been partially cleared of boulders during the Depression, but they were mainly boulder-strewn river terraces covered in bracken and regrowth tōtara.'

After unsaddling our horses and leading them back to the house paddock, Barry and I went inside for lunch while Tim carried on with his work. Once we'd eaten, we settled on comfy chairs and admired the view and Barry told me how, after getting married in 1961, he and Sue had built this house. It sat just above the river, where it curved in like a pipi shell, and the surrounding hills had reclaimed the bush that had been cleared a century ago, and tūī sang their bell-ringing chorus. It might have been tough land, but I could see why you wouldn't want to leave this place.

For his final years of schooling, Barry had been sent to board at Wellington College but, he told me, 'There was never any doubt about me coming back.' And, at the age of 18, that's exactly what he did. 'I finished up there on the Friday, and started work back here on the farm on the following Monday,' he said. 'No, this was the place for me.'

It was 1953, and his father had converted the farm to dairy after the wool price had dropped. So Barry slotted into the cowshed, milking, feeding calves and doing general farm work. But he never really liked dairy farming, so instead he found work as a shepherd, mustering on the Arcuses' big block up the top of the gorge. It was wild, untamed country, flush with pigs and deer. 'I got up there and I thought, Man alive, this is the life for me,' Barry told me. 'It caught my heart, I suppose you could say. It was good mustering country, and it was where I belonged, with my dogs and my

It feels a world away, but the bustling town of Ōtaki is a stone's throw away from Tiroroa and beyond that is Wellington.

Tim and his horse sit back and let the dogs do the work for a moment.

horses. I never thought about doing anything else, and Tim has been the same.'

By this time, Tim had come in too. Overhearing his dad, he nodded. 'This is my happy place too,' he said. 'What else can I do? Give me dogs and horses, and I am a happy man. Dad and I have that in common.'

Tim also swore by using horses to muster. 'It just works with dogs and horses together. It makes sense. Dogs get bike happy when you go everywhere on the quad. They know you have to concentrate on what you're doing, and then they start pushing the boundaries and being silly buggers.'

Now a father himself, Tim wasn't certain about what the future would hold for Tiroroa – his daughters had their own lives outside of the farm – but for now he was determined to keep his family's legacy going as long as he could. Still, even this last bastion of an older way wasn't immune to the encroaching influences of the outside world. When government subsidies were removed in the 1980s and income became tight, the family had been forced to downsize their stock – as was the Arcus family – and bring it off the wild country up the top of the gorge. The last big muster there was in 1997, and after 80 years of being farmed, the land had reverted to fern and scrub.

Barry still remembered mustering up the top of the gorge with the Arcus family: 'It was always a challenge, because the weather was so unpredictable.' The stock, he said, were hard to find and fossick out in the bushy gullies, and had a habit of putting themselves up on a tricky-to-reach ridge. 'But they were enjoyable times. It was the end of an era and a whole way of life after that last muster.'

*

This slice of land in the Ōtaki Gorge was an open window into the area's past, and visiting it gave me a real sense of how much farming communities can change within a century. How horses – once so revered – can become bookmarked in the 'what was' chapter. Here, horses were once used in many aspects of daily life, and before the Pākehā settlers came, Māori had used their hōiho to traverse the terrain, which they held in the highest regard. But, like so many areas in the north once famous for their horseback mustering, this area had largely got rid of its horses and exchanged them for quad bikes.

Tiroroa was the last man standing. The Mansell family, in their corner, remained determined to uphold their history, to keep their passion for horses and their place on the hills alive. On this farm at least, horses still had a job to do and the shed remained full of saddles and bridles. The past was not lost here. It was honoured every day, by Barry – who remembered it like it was yesterday – and by Tim, who carried on farming his handkerchief farm even as the land around him was progressively sold up. But, I had to wonder, how long for?

5
IN THE BLOOD

Ruanui Station, Mataroa, Rangitikei

It was so early it was still dark when I arrived at Ruanui in the back country of Taihape, up where the hills are layered like a rumpled sheet and stretch as far as the eye can see. There, I found a cluster of men gathered outside an ancient but well-kept stable block. They were outnumbered by dogs, and in the yard five horses were tied up. They broke their huddle to greet me, head shepherd Ricc leading the charge with a firm handshake and a bright grin, followed by shepherd Max and junior shepherd Matt. Finally, the boss, Jeremy: 'Morning,' he said, a puff of frosted breath escaping with the word.

There's always something magic about these pre-dawn starts, with the sun still yet to stir and the whole day hanging in the air. Anticipation of the day's work filled the air around the team and I breathed it in too. I couldn't wait to get started.

Once duties were assigned, the prep and saddle-up commenced. 'The horse I'd thought you'd ride is lame,' Jeremy told me over his shoulder, as he hustled about getting people, horses and dogs sorted, 'so you've got Covid instead.' Then he stopped, and added with a grin, 'The horse, not the pandemic.'

Covid was a little dun mare and got her name from the fact she'd been a break-in project during the country's first lockdown. 'She doesn't do anything nasty,' Jeremy added. 'She can just be a bit hot.'

I caught a smile from Matt, who was saddling up alongside me, his shaggy mullet sticking up in all directions. Fresh out of school and 17 years old, he was learning the ropes and a bit of horsemanship under Jeremy's easy-going and experienced watch.

'And you'll need this out here, too,' Jeremy said, handing me a breastplate – a leather strap that goes round the horse's chest, connecting to the saddle and helping to keep it in place when the going gets up and down. I hadn't seen much flat land since leaving Mataroa, and over our backs rose the distinctive green Taihape hills.

Covid is a born, bred and trained Ruanui Station horse.

Taking advantage of a high point to look for cattle.

As we all mounted up, Covid gave me the little skitter I'd anticipated, but I slotted into the group of keen musterers, and together we hotfooted it uphill. Our horses were fresh and eager to get on, heads held high and reins kept short. Sitting at 550 metres above sea level, Ruanui was 3130 hectares of volcanic hills. Not the rolling variety — more your straight-up-and-down sort. The guts were deep and the highs looked up towards Mount Ruapehu, the tribal maunga of Ngāti Rangi, which was just visible through the gloom. By now, the sun had stirred, but its light barely reached us. Instead, a cold drizzle tagged along.

I followed Jeremy, whose wide-brimmed hat and stock whip seemed to be a permanent feature, while the others took different routes out to where the cows and calves were. I could feel Covid's nerves running through the reins and into my hands, and I tried to relax so she would too. Horses have a highly developed perceptive awareness of their environment — they are prey animals, after all — and Covid had all of her senses switched on. When people say horses can smell your fear, it's true; they also react to a rise in your heart rate, making them good mirrors of stress, anxiety, fear and anger. Covid was figuring me out, just as I was her.

'She'll probably like having a softer rider on her, I'd say,' said Jeremy, noticing Covid's head relaxing a little. 'Horses know, eh? They respond to how you treat them. And they make you think about what you're doing. Get on a horse mad, and you're bound to have a tricky ride.'

In response, Covid twisted her jaw in a huge yawn – a sure sign in a horse that it's relaxing – and Jeremy and I both laughed.

Jeremy's mare was also young and green, not long broken in and therefore still getting the hang of things. 'And what better way to do it than getting her out into these hills?' Jeremy said as we reached the top of that first big chug, horses puffing, sweat glistening.

Jeremy and his wife, Jo, had managed the station for the last decade, and its third-generation owners, Meredith and Andrew, were supportive of them using horses for the daily work. 'I wouldn't want it any other way,' said Jeremy. 'Horses are perfect for the hills out here. You work the stock differently when you're on a horse.'

They bred their own horses on the station, sticking with strong, well-boned mares and a station-bred stallion. Breaking-in time was a chance for Jeremy to pass on his knowledge to Ruanui's young shepherds, and he used the horses as a way to impart a bit of life learning, too. 'A horse needs looking after,' he explained. 'It's a responsibility. You have to brush it, feed it, care about it and think about what it needs. That's a good thing for a young person to have to do.'

He'd gained this knowledge in the same way he was passing on, by picking it up from those he'd worked for and with. One of those first mentors was his dad. 'I never wanted to be left behind when Dad went mustering. We spent a lot of time on our horses, and all us kids would go off and have adventures. It was just what we did, and I guess it was what I would always end up doing.'

For his part, Jeremy preferred the firm-but-kind approach. 'It's just a matter of understanding horses,' he said. 'Each one is different, and you adjust what you do to suit them. There's no use being hard as nails. You'll end up with a horse that is too.' Horses, he explained were a necessary tool on the steep country he managed, but they were also your mate. 'There's nothing better than a good horse and a sunny day,' he said.

*

Once the cattle are in front of the musterers, it's a matter of keeping them quiet and controlled.

Keeping the cattle together and moving towards the yards for the afternoon's work.

We were riding down a steep hill into a tight gully, leaning back in our saddles and making our own zigzag path to the bottom. On the far side of the gully waited a tight knot of cows, and there was Max, way up on the other shoulder, silhouetted against fence and sky. Jeremy radioed him to tell him to bark up a dog and encourage the cows to move down, and shortly afterwards a staccato of huntaway barks echoed towards us. The cows responded, nudging their calves forward and away from the dogs. Max hadn't had a farmy upbringing, so hadn't ridden before coming to Ruanui a few years before. When I'd asked him earlier how he found working with horses, he told me he enjoyed it. 'It's better than walking, that's for sure,' he'd said.

Jeremy and I started working another mob of cows down to the others. The usual and well-practised gather-and-push kept us busy, and eventually Max met up with us, joining his mob of cows in with ours. Then, as Jeremy hopped off to open the next gate, he warned me that things could get a bit interesting in this part of the station: the track went two ways round the bottom of a conical hill, he explained. 'And, if we aren't careful, the cows can end up back here, right where we started.'

Just then, the clattering of a horse's hoofs alerted us to an incoming rider, and Jeremy opened the gate a bit wider, a big grin breaking out on his face. 'She's coming in hot!' he called as Jo came to a dust-cloud stop on her chestnut horse right behind us. After dropping the kids at school, she'd speedily saddled up and high-tailed it out here, arriving just in the nick of time. 'She always turns up when we need her,' said Jeremy.

Jo's horse was a bit different from your usual station horse. 'He looks a bit posh, Jo!' I said, and she laughed. 'Yeah, he is,' she admitted. 'He was meant to be a dressage horse, but he didn't work out and was going to be sent to the knackers yard. So I got him.'

Within moments, Jo was hooning here, there and everywhere, keeping the mob of cows tight. She was a great rider, and used to be a gun barrel racer. Like Jeremy, she'd been brought up with horses and, once she was in the saddle, transformed from a quietly spoken farmer into an assertive kick-arse who got the job done. It was a joy to watch her riding, especially because she seemed to know what the cows were going to do before they did it. When I told her this, she laughed, clearly a bit embarrassed. 'It's just that I have spent my life doing this,' she said, downplaying things. 'It's nothing special. It's just what I know.'

Jo and Jeremy were both Taihape born and bred, and neither had ever strayed far. Ruanui was home for them and their three kids, and it was much more than a job. It was their day-in, day-out, all-encompassing way of life. Living this way, with the horses and the hills, was generational for Jo and Jeremy. As Jo said, 'It's what we know deep in our bones.'

*

As calves started to get mis-mothered and the cows began to get toey with the dogs, the energy began to ramp up. Barks, moos and all of us shouting and hollering made for a chaotic soundscape. Dust was billowing, horses were sweating, and coats were stuffed into saddle bags. Individual cows peeled off from the bellowing mob, defending their calves and fighting the dogs. And, once one turned, others followed, setting off a domino effect of cows that hurtled towards us, past us, and way into the distance.

The chase was on. We rode as fast as we could, jumping ditches and bogs and negotiating the sides of hills at speed, generally doing whatever we could to get in front of the rampaging cows and calves. Adrenaline-fuelled, we rode

Rylee and Ponga in the thick of it back at the yards.

Things heat up as the cows face up to the dogs. Rylee helps his dad out on his pony Ponga.

with instinct, caution thrown to the wind. Covid responded with gusto; she knew exactly what she was meant to do, and she did it fast. Everyone was riding hard, everything concentrated on the task at hand.

Once we got the cows turned, we began again, pushing with all that we had – our horses, our dogs, our sticks, our voices and with the ear-piercing crack of Jeremy's whip. He flung it high in the air, and with a flick of the wrist cracked it. His horse didn't even flinch – all of the horses at Ruanui soon got used to having a whip cracked from their back. It was an effective tool to get the cows moving, plus it gave the muster that feel of the Wild West.

We were all focused on one thing: moving forward without letting the cows get past us. But they did, of course, and over and over we started again. We lost the battle one minute and won an inch the next. We kept going. The yards and morning tea were the end goal, and we had no choice but to get there. It was absolute teamwork, between the horse and rider, dog and master, all of us helping each other.

This was how things had been done on this land for many years. Horses had been intertwined with Ruanui's history since the earliest Europeans made pushes into the region. William Colenso had been the first to travel into the area of Taihape, but the Reverend Richard Taylor was the man who ventured into the more specific area of Ruanui in 1860. Alongside 100 Māori, he had journeyed up the Whanganui River, across the Waimarino Plain and into the nearby Turakina Valley. The group, Taylor recorded in his diary, 'reached a little kāinga called Ruanui' and later the settlement was reached through discussions at Kokako, a pā located near the Hautapu River. It was later reported in the *Wanganui Herald* that the government had not acted fairly towards Māori when purchasing Ruanui: 'I am certain if anybody other than the government had bought this land the natives would have very much been the gainers,' a columnist named The Rambler wrote.

John Studholme and his wife, Lucy, were the first settlers to lay a claim on Ruanui in 1873, and their son Joseph followed them in 1890. Cattle were the first stock to be brought in, but that proved tricky, with many of them 'going bush'. Sheep were more successful, and their wool was taken to Napier using a large mob of packhorses. The going was tough, with no dray road within miles of the station, making it isolated from the small settlement of Taihape, 12 kilometres away. The wool was first taken out to Turangarere for scouring and repacking by as many as 30 packhorses loaded with two quarter bales of wool each. The next part of the journey was made with around 70 strong-horse-drawn wagons to Napier, making the whole enterprise a 160-kilometre, three-week-long trip, which was taken in all weathers and made with no properly established roads.

Those same strong horses that carried the woollen cargo through rough terrain and across rivers later carried soldiers in the First Boer War. In a letter to the mayor of Whanganui in 1900, J.F. Studholme offered up the services of 'five Ruanui men and their horses', and declared that while his horses 'were not much to look at', they were very much suited to war. 'The horses are well used to hilly and rough country, and finding their own feed,' he wrote. 'I consider that the third contingent, if composed of men accustomed to rough back-country life and being good riders and moderate shots, will be invaluable.'

The Studholmes became established as horse-breeders with their well-regarded stallion Recluse, and they would bring over 40 horses – 'hacks and harness horses all broken to saddle' – to the Feilding Saleyards annually. The family was also very social, and as well as hosting cricket matches and balls, entertained many high-profile guests of the time.

Rylee and his pony listen in on cattle talk between Jeremy and Ricc.

A stand around catch up in the yards is necessary after a long day's work.

Lady Glasgow rode in on horseback for a visit, something that was reported at the time in the *Wanganui Herald* as being quite exceptional, as she was not known to be an avid rider. Such was the pull of Ruanui and its horses.

George Carpenter, Andrew's grandfather, bought the station in 1952, beginning the current and ongoing chapter of Ruanui's history. George carried on the legacy of horses, as did Andrew. And Jeremy, as the current manager, was a modern-day cowboy who had intentionally followed horses and fitted right into the horse-dominated history. In his early days as a shepherd, he'd expressly chosen jobs on farms that didn't mind him bringing his horse. 'I guess there are a lot of places that don't use horses anymore, and that's a real shame. We are lucky that we can here.' Both he and Jo took real pride out of the fact that they still used horses at Ruanui daily. 'It would be a sad thing to lose,' said Jeremy. 'These hills shouldn't be without horses. This is horse country out here. It's in the history.'

*

An escapee calf had Jeremy and his horse high-tailing it into the distance, chasing and turning it back into the fold where it found its mum. As the mob stitched together a bit firmer, we rode on and had a chance to catch our breath and talk.

As we passed a large slip on a hillside, Jeremy drew my attention to it. 'The weather has thrown a lot at us lately,' he said, 'but at least we can just keep going, because of the horses. Some of the stations that have stopped using horses might just be regretting that decision. You can ride a horse through a slip and go places a bike can't.'

Once all the cattle had been brought into the yards, and the horses were tied up in the shade, it was time for smoko up at the shepherds' quarters. Over a delicious chocolate log with real cream – whipped up by the station's chef, also called Jo – and a cuppa, Jeremy outlined the plan for the afternoon. 'We'll be draughting the babies from their mums so that we can ear-tag them, give them a vaccination and castrate the boys,' he said.

Here at Ruanui, the motto was 'if you can do it on a horse, you do', and the afternoon proved no exception. First, the draughting: Jeremy used his nimble mare to get in between each mum and her calf, pushing them into separate yards, while I assisted on foot at the gate. Low on its strong back haunches, the mare bounded forward, spinning and shifting her weight efficiently. It was quite something, and

Jeremy takes the high road, pushing the cattle to a new paddock after the yard work.

Rylee takes the low road, mirroring his dad.

with ears pinned back this horse meant business. The cows respected the horse much more than they did a human on the ground, and swiftly and surely we had a mob of bewildered calves in one yard and their bellowing mums in the next. It was a much safer, quicker and frankly more fun way to get the job done – much better than having a whole team of humans running around in a flap among the anxious cows.

Then, once the dust settled, we cracked on with the rest of our duties, and slipped into the steady rhythm of monotonous hour-on-hour work. The music got cranked up and the familiar banter was bandied about, while dogs and horses slept and calves mooed relentlessly for their mums.

Just in time to take the cows back out, Jeremy and Jo's 12-year-old son Rylee appeared on his grey pony, Ponga. With his stock whip in hand, he fell right into step with his dad. The father-and-son duo made quite the sight. Jeremy's horse was double the size of little Ponga, but the pint-sized pony kept up with jaunty gusto, Rylee weaving him back and forth. This team had it so thoroughly under control that I simply eased back and watched. But, once we started pushing the cows down the road, I rode up to Rylee and asked him about his pony. 'She was quite naughty when I first got her,' he said. 'And a bit head shy.' But he took his time, he said, 'And I got her right in the end. My little sister will be able to ride her now.' He was clearly proud of this.

'What do you like most about horses?' I asked him.

'Everything,' he said, grinning shyly up at me.

*

My visit to Ruanui gave me a real sense of what it meant to live alongside horses, and a glimpse of what it meant to pass on the passion for them to the next generation – not just your own children, but to your juniors on the farm. The best and only way to learn to muster on horseback was to get in the saddle, and it was inspiring to visit a farm with a focus on the future and a determination to keep young shepherds and young riders coming through, so that knowledge wouldn't be lost. And, as I was about to discover, Ruanui was far from the only place where the goal was getting through the musters to come, rather than hoping each muster wasn't going to be the last.

6
PRETTY MAGIC

Smedley Station and Cadet Training Farm, Tikokino, Hawke's Bay

Wooden rails were made to be leant over, and that's exactly what I was doing one misty morning at Smedley Station while second-year cadet Tyrone slowly tightened the girth of his tooled-leather stock saddle. As he did so, he murmured reassuringly to his dark chestnut mare, Āio – the Māori word for 'be calm, be at peace'. She shifted, and so did he. A pause while she turned her head to nuzzle at the saddle.

This was Tyrone's final year at Smedley, and he was one of four lucky students chosen to break in a horse. This was seen as an honour, one not taken lightly – Tyrone and his peers were spending every day that week working with their young horses, getting them used to the halter, to being led and tied up, to having their feet lifted, and to having a bridle, saddle and breastplate put on. Each stage had to be solid before the cadets could put a foot in the stirrup, and each stage took time and patience, repetition and quiet perseverance. These horses were fresh, and there was no room for bravado or anger.

Watching Tyrone with the young horse, it was clear he had the instinct. The thing that wasn't always there but, when it was, could be quite something. It was in the way he moved, knowing when to drop a shoulder and his gaze, knowing when to strengthen his stance and when to soften it. He paid attention to the way the horse responded and reacted, and he kept the channels of communication open.

'He's a good horseman,' said Rob, the station manager, a shorts-all-year-round kind of man who was tall enough to see easily over the top rail, while I had to stand on the second rung just to get a peek in. 'There's always a stand-out cadet with the horses, and this one is it.'

'I think it's time to get on,' Tyrone said, looking back at us. We threw thumbs-ups and words of encouragement his way and, after tightening the chin strap on his helmet, he stood with his hands on his hips, taking it all in for a moment. Deep breath in, deep breath out, then a calming

Two young Smedley horses in training.

Smedley cadet Tyrone has a natural way with horses.

stroke of the mare's tightly packed neck muscles. Tyrone's stirrups were long and his reins short as he eased himself up into the saddle, softly, softly, and as slowly as he could. The mare stood firm, and her ears flicked back and forth, nostrils puffing. Tyrone gave her another pat and a murmur: 'We have all day, Āio.' She shifted her weight, getting used to his, and all the while there were those ears – back and forth, alert and aware. The little mare settled and, with reins loosened a notch, Tyrone applied the subtlest amount of leg pressure. Āio moved forward, and Tyrone let her, adjusting his weight and legs so that she turned in a tight circle.

At the rails, Rob and I let out our breath and exchanged a smile. What we were watching was a bit magic. A communion and a beginning.

*

Horses had always been a part of life on 5054-hectare Smedley Station, which sat just out of Tikokino and rested under the Ruahine Range. When the station was first taken up in the 1850s, horses were the only way to get in and out of it, and to traverse its hills.

Smedley got its name from the Lancashire hometown of its first European owner, Josiah Howard, who had been from a well-off family of silk millers but had come to Aotearoa for tragic reasons. On his first wedding anniversary, his wife and child both died, and he was so devastated he swore never to marry again and that he would leave his homeland forever. He first tried his luck (and lost it) on the Australian goldfields, then went to Central Otago, and finally headed north to Ahuriri/Napier.

At the Government Land Office, Josiah peered at a map of Hawke's Bay, and there among the blocks not yet taken up, his gaze had settled on land north-west of Waipawa – an area where Ngāti Te Upokoiri and Ngāti Hinemanu had built kāinga in the foothills, before European settlers arrived and started clearing the forest en masse. Josiah had exactly zero farming experience, and he picked the block that would become Smedley at random. And, once the deal was done, he travelled up the Waipawa River alone, and was only joined later by his brother James, then his nephew Robert.

When Josiah eventually died in 1919, he left the by-then completely debt-free property to the Crown, with just one stipulation: it was to be used to train young farmers in all aspects of station life. But the land was in a sorry state. When Josiah's health had first started to slide back in the 1890s, so too had the station. Scrub had got a foot in, stock numbers were drastically cut, and fence-lines fell into disrepair. So, when the Crown came to inspect their gifted station, they deemed it too rundown to be a training farm. It was put in the too-hard basket … until a good old public outcry came to the rescue, and the Crown was forced to reassess. The station was subsequently passed to the Public Trust, and in 1931 – 12 years after Josiah's death – the first five cadets came on to the station to begin their training.

And, almost a century on, Smedley was still a permanent place of learning. Here, education was free for the 13 cadets selected to leave their boots at the hostel door each year. No matter what their prior experience, all of the cadets used horses for stock work. As Rob told me, it was part of the tradition of the land. 'Using horses is something that we want to see continue,' he said. 'It's all part of teaching good stockmanship, respect for animals and discipline. Horses are an important part of this place.'

*

The stable walls of Smedley are covered in historical messages from past cadets.

A Smedley horse waits patiently for the day's work to begin.

'She's a nice horse, that one,' said Jordan, a senior cadet who'd come to join us at the rails. Jordan was from Raetihi, and had been brought up around horses. He came from a solid rural upbringing, but that wasn't necessarily the case for all the cadets. Alongside him stood Josh, who was from Auckland; for him, horses were just one of the big learning curves Smedley was dishing up. 'I didn't know the front end of a horse from the back when I was first here,' he told me. 'And now here I am, breaking in a horse! It's cool and I want to keep doing it.'

Attitude was more important than rural birthright here at Smedley. After all, it's not like Josiah Howard had any experience when he first came here. According to him, 'You must find out whether these young people have the aptitude, energy and ambition. If they have not, Smedley might as well be planted in a garden.'

Rob, who'd once been a Smedley student himself, echoed this sentiment. 'My theory is that you can teach a good person anything, but sometimes it's harder to do it the other way round,' he told me. 'The cadets that we take on have to be passionate about farming, and they have to want to learn.'

The Smedley cadetships were highly sought-after, and there was a bit of a tendency for alumni like Rob to find themselves back at the station. Rob told me he'd never really thought he'd come back, but when he did – many years later – he remembered how much he loved this place. 'And it hasn't changed,' he said. 'It looks just like it did when I was here as a cadet.'

It was an incredibly beautiful stretch of land. The rolling hills, covered with old stands of tōtara, beech and kahikatea, met with higher peaks, bush-clad and topped with mist. It was parklike, almost as if the hills had been landscaped with an artist's eye. I could see why you'd be drawn back here. This was a place full of history, of meaning, where young people had been coming for decades to learn to work with horses, and in the process gaining greater life lessons, too. 'They can't lose their heads. The horses will react, and things will go wrong,' Rob explained. 'These cadets have to put their egos aside, and they need to listen.'

Calm, patience and perseverance: all qualities the cadets were honing that would go on to serve them throughout their lives, in all kinds of ways. 'If we instil this into them now, then they can carry it all with them when they leave,' Rob said. 'Gives you a bit of hope for the future, eh?'

*

Tyrone was back on the ground now, taking Āio's saddle off. 'That's given you enough to think about for now, eh, girl?' he said.

'Do you think she'll be ready for lambing?' I asked him.

'For sure,' he replied. 'She's doing great. A special horse, this one. I'm not going to want to give her back.' He winked at Paul Wagstaff, the station's stock manager, who'd come over to check on progress. Paul just chuckled, a good-natured man with a solid sense of humour – which I can only imagine he needed, as one of the main points of contact for the cadets.

Tyrone and the others tasked with breaking in the new horses were aiming to have them ready for the oncoming spring. When lambing started in a month's time, these young shepherds in training would be out on their fresh horses, checking the ewes and lambs every day for about three weeks.

Not many still did the lambing beat in this way. It was a huge commitment to check the stock daily, and it was possible these cadets might never do it so intensively again – but, as Paul said, 'This is a place for them to learn, and to have that responsibility and knowledge. It's a tradition they can take with them.'

Smedley cadets head out on a morning lambing beat.

A sketch of a favourite dog done by a Smedley Station cadet.

And horses were the perfect vehicle for the job. 'We don't want the ewes and lambs disturbed, so going quietly on a horse is the only way to do it, really,' Rob told me matter-of-factly. He was as straight-up as they come, not at all romantic about things; to him, horses were a practical tool for accessing stock. 'You just see and notice more,' he said. 'It makes sense. Horses are the way to go out here, no question about it.'

'Actually, Carly,' Paul said, turning to me, 'why don't you come back and do the lambing beat with the cadets?'

*

When I returned to Smedley a month later, I stuck out like a sore thumb among the young cadets, but I hastily saddled up my horse: a big golden palomino called Oakley. All the horses here were Smedley-bred, chunky without being too heavy-set, nice and steady. 'We need some good quiet ones to put the juniors on who have never ridden,' Rob told me.

Once we set off, I soon fell into step with Aliyah, who was up on a chestnut horse brushed to a coppery shine. The only girl in the junior year, she was from Golden Bay, and had grown up in the saddle. When I asked her about life at Smedley, she admitted it had been hard having no other girls around, 'but the boys are all really good to me and I am learning a lot.' And, while it was true that she was heading into a male-dominated industry, that wasn't to say there weren't any women who were shepherds out there. There were, in fact, quite a few – 'and good ones at that,' I reassured her, thinking of Alice, Kelsey and Bridget from up the Awatere.

We crossed a creek, and Oakley nudged his way up to the front of the group. He hadn't been out in a while, so he was a bit full of himself, putting on a little prance now and again. I noticed the seniors taking the lead, both on their horses and also of the group. This was how things rolled at Smedley. Each senior was assigned a junior, and acted as their mentor and told them what to do; this gave the seniors an early taste of shouldering some responsibility. Additionally, it was only the seniors who had dogs at their horses' heels. As with everything here at Smedley, there was a rule behind this: once you were in your second year, you could have a dog of your own. First, a huntaway, then – once you'd proved you'd done a good job of training it – you could get a second, a heading dog. This meant the first-year cadets had to learn to muster without dogs. 'And that's a good skill to have,' Paul told me. I nodded, thinking about my own no-dog mustering journey so far and all the little tricks I'd had to learn to compensate.

One senior who I could see had gained respect from the others was Connor. Like Josh, he'd also come from Auckland and wasn't from a farming family. It was a big step up, he told me, going from following to leading in your second year. 'We learn a lot this way. How to communicate and manage people. Knowing when to push, and when to ease off a bit.'

*

Jordan, the senior cadet from Raetihi, paired me up with first-year cadet Sam, who was on a strong little grey and clearly still finding his way round a horse – but he had a big grin on his face all the same.

'I'll take the low end of this paddock,' Jordan said. 'You two take the high end.'

Sam and I nodded dutifully, and did as instructed. Our job was to check on the pregnant ewes and on the lambs that were just starting to be born. We made our way quietly around the stock, careful not to cause any disturbance, and

Oakley is a Smedley Station bred Palomino with plenty of bounce.

the sheep hardly noticed us. It would have been tricky to be so unintrusive on a quad bike. Loud machines and people trigger a fear response in stock, but this wasn't the case with our soft-stepping horses. What's more, on horseback we were able go off-track and take out-of-the-way routes to weave around the ewes and give them space.

'It's a nice way to do things,' Sam told me. 'I hadn't realised before how different it could be using horses. You see more, and you can take your time.' His words were, unbeknown to him, a direct echo of Rob's from my first visit to Smedley. The theme here was strong, and I found myself wondering how it could be applied more broadly. What would it mean for our world if we all paid closer attention, and took our time to do things carefully and properly?

We paused for a moment to scan the hills. Now that I was out here actually on the station, away from the hostel and the houses, I saw just how special this place was. The corner we were in was rolling and green, and native tōtara stood nearby, mature and mighty. They were quite a sight, especially knowing how many tōtara were cut down when colonial settlers first cleared the land in the 1800s. I felt I was being granted a glimpse of what could have been: these trees spoke of what the land had been once, and what might have remained. And there was more to these trees than just beauty and awe; they were practical, too, providing the perfect shelter for lambing. Here was a place that ewes could hunker down when the weather turned nasty, and as we rode through a grove of tōtara, ewes and lambs peered out of the hidey-holes formed by twisted tree roots.

Once we were done, we joined back up with the rest of the team, and I rode back to headquarters with Ben, a first-year cadet from Pahiatua. He was riding my pick of the horses, a dark chestnut with a good dose of quarter-horse blood called Tane. Ben was in his element on Smedley. 'My family is horsey, and Mum has horses at home. This is where I like to be.' He was a rodeo rider, he told me, and like many of the cadets he was keen to find work after Smedley, somewhere that horses were on offer. I'd heard so many stories of stations getting rid of horses – because of economics, because of time management, because of the 'extra hassle' – that my heart skipped a beat every time a cadet told me they wanted to carry on working with horses in the future. Here, horses were not a thing of the past; they were a way forward.

*

I slept soundly that night in the old shearers' quarters, next to a crackling and roaring fire, and was woken at the crack of dawn by the sound of horses moving very close to where I lay. Some of the cadets were already out in the yard, rounding up the horses and getting ready for the day. Talk about learning a good work ethic.

After breakfast, we were straight back into the routine of the lambing beat. I jumped into a ute with Tom, and we set off for Back Whare to meet up with some senior cadets who were doing the back-of-the-station beat. Tom coached the cadets on their theory, and had the big hands and friendly, weathered face of a man who had worked the land for many years. He had a deep care for Smedley, and for his work here. 'I can pass things on and help the next generation here,' he said, 'and I love that.'

Yet again, I was the designated gate-opener, but after I'd jumped out and opened the eighth one, I started to lose count. 'And you're going to have to do these all on your own on the way back, Tom!' I said. He chuckled. 'Don't worry, I'm well used to these gates,' he replied. As we made our way towards the whare, the country got bigger, the gentle hills turning into rocky outcrops and bush. The country was harder out here, thick with mānuka, and fell under the dark gaze of the Ruahine Range. On this particular morning, a low mist hung over everything, adding extra moodiness.

Finally, pulling up at the bottom of a sweeping valley cut through by the Makaroro Stream, we reached the hut, where we were greeted warmly by Jordan and Callum. Dating back to the mid-1950s, the hut looked like a natural extension of the land, all straight lines with two big windows and a smoking chimney. And, once inside, I realised it held a luxury I had not encountered in a mustering hut yet: a bath! The ancient old clawfoot bath stood proudly inside, and had its own potbelly for heating the water. A blissful thing, I bet, after a long day out on horseback.

The walls of the whare were etched with drawings done by cadets over the years that told of horses, dogs and camaraderie. It was a highlight for the seniors to stay out here, and they took turns doing it during the lambing beat, riding their horses out at the start of their stay. While here, they were responsible for checking the back reaches of the station and feeding out a mob of cows. 'Plus we do a bit of general work – firewood, fencing and tidy-up jobs,' Callum told me.

Outside, tied to the fence along with three paint horses, I noticed Āio, the little chestnut that Tyrone had been

Many meals have been cooked by Smedley cadets on this old wood stove.

Callum and Jordan in front of the Back Whare.

breaking in when I first visited Smedley. 'Yeah, Tyrone was up here, but he had to make an urgent dash to the vets early this morning with a dog with a twisted gut,' Jordan told me.

I asked how Āio was getting on.

'She's an amazing little horse,' Jordan replied. 'Tyrone has done a good job. She follows him around like a dog.'

My ride for the day was Vegas. 'She's very cool,' said Callum. 'Everyone likes riding her. She knows what she is doing.'

The moment I got on Vegas, I added myself to the list of people who loved her. Sometimes there's just something extra special about a horse. They might have a sparkle of personality that sets them apart, or they just seem to be particularly switched on. Vegas had both, and the most magnificent mane to boot.

We headed out towards the big country of Smedley, Callum on Bandit in front. He was from the neighbouring town of Waipukurau, so this was basically home turf for him – and that's how he liked it. 'This is where I want to be,' he told me as we passed through Dougie's Block. The going was wilder out this way, with steep climbs that made the horses puff. We stood up in our stirrups to make the going easier for Vegas and Bandit; this was something that Paul insisted the boys do. 'It's now become a habit,' Callum said.

The wairua of Smedley was potent: the lineage of learning that the land had provided, and the respect it had been given imbued it with a certain feeling of mana. This land was held up by all that had come before, and I could see that Callum felt a real pride to be showing me around. 'It's pretty magic, eh?' he said, as we paused at a high point.

'It is,' I replied.

7
THE REAL DEAL

Mangaheia Station, Ūawa/Tolaga Bay, East Coast

My friend Lindy and I cruised along in her beat-up Toyota with a giant huntaway called Bully in the boot, an old heading dog called Skip asleep on the back seat, and Lindy's two teenage moko squished in among bags, coats, boots, trays of eggs and dog food. All legs, Thorston was contorted in the front seat around our various belongings, while in the back seat Anika was all smiles in spite of the old dog farts she reckoned Skip was inflicting on her. We were en route to Mangaheia Station, where Lindy's son, Leo, was the manager.

Two years earlier, when the station changed hands, locals had feared it might follow the same path as many of the area's other big stations, which had relinquished hard-to-farm hills to trees. Recent flooding in the area had brought intensive planting, and its consequences, to the fore – the scenes of slash strewn like giant matchsticks along the coastline, clogging up waterways and, in places, taking out entire bridges, had been an eye-opener and a shout for tighter processes to be put into place. But Mangaheia's new owners had kept it running as it always had been, and it remained a classic East Coast hill station, where sheep and cattle still grazed the hills. 'It's a great place for young shepherds keen on horses,' Lindy told me as we bumped along the road.

As we drew closer, the hills did too. They hummed a different tune from the mountains down south, and the further north you went on this coast, the wilder it became. This was Ngāti Porou and Te Aitanga-a-Hauiti country, and the rhythm was ancestral. Goats grazed the side of the road, big old dreadlocked billies watching over their harems, and the going was potholed and rough – since severe weather was thrown this way regularly, the road was always in a state of disrepair, and often closed. Slips scarred the landscape, and whenever big rains fell, the ocean-neighbouring settlements got cut off from the outside world.

To live here required tenacity, and horses. Among other things, a horse was a handy means of transport that didn't require trips to town for petrol, and the East Coast was the domain of gritty people and legendary horses. Out here, where the crashing sea was never too far away, you held on to your traditions. Whakapapa lines and knowledge flowed naturally. And horses were part of the landscape, ridden by pretty much everyone who called this whenua home. Horses stood proudly on front lawns and, bareback with a halter, provided a means of getting around the cruisy little bay. They were a natural part of life.

*

It was a bold and early 5am start the following morning, because we needed to make sure the station's huge and high paddocks were mustered up before the docking crew arrived at the yards at 7.30am. I jumped into the ute with Leo and his 12-year-old son, William, who was tagging along with the shepherds that morning – kind of like a very extreme bring-your-kid-to-work day – and, in utter darkness, not a star in the sky, we tore down to the yards.

There, I got a hasty introduction to the shepherds: George aka Sheildsy, Finn and Tom. Then Leo pointed me to my horse for the day.

'What's her name?' I asked.

'I just call her the chestnut horse,' Leo replied, as he mounted his big black horse. 'She doesn't really have a name.'

The chestnut horse was all muscle. I climbed into the saddle, and then we were off. There were no head torches. The shepherds didn't use them. Instead, they relied on their knowledge of the land and their horses' superior eyesight. I relied on the white patch on Tom's horse's butt and the intermittent flashes from the dogs' shock collars. The boys were fast on their horses, walking out wide and trotting where they could, on a mission to get the job done, with their dogs at the horses' heels. It was all I could do to keep up. But I couldn't lag behind or I'd get lost. Shit, I thought, this is as hard as it gets. But, as terrifying as riding along in the pitch dark was, it was also totally exhilarating. It felt like the whole world was asleep, except for us.

My chestnut horse started to huff and puff as we began to climb, and when we reached the top of our ascent the sun was just starting to shed some light on our situation. Behind and before us stretched high hills, and the orange glow illuminated craggy heights and a gleaming pond. It probably wasn't much past 6am. This landscape felt like a gift – but there wasn't much time for appreciating it, because we were already off again, flinging reins over horses' heads and jumping off to hotfoot it down into a valley. It was steep and muddy, still pretty dark, and the boys were like mountain goats. I was not. I slid down on my butt, falling on my arms and grateful the dark and my place at the back meant no one was witnessing how incredibly uncool I looked.

It was do or die around here. No one, not even the dogs, wanted to be late with getting the sheep in. At Sheildsy's command, the dogs were off like a shot and everyone got to work. Ewes with their lambs are not the easiest animals to muster, and these were no exception. They were hard-nosed, not keen on budging, protective of their young and much braver with the dogs than usual. If you pushed too fast, the lambs got mis-mothered. And they were prone to scattering, which then gave the ewes a chance to gap it, shooting off erratically in all directions.

We slowly moved the mob forward, over creeks, and managed to get them into the yards just as the docking crew turned up on two buggies pulling trailers chocka with gear. They were late, but we didn't mind one bit – at least they hadn't been waiting for us, or watching while we had a bit of trouble at the last leg.

Leo surveys the land below, watching his team of shepherds in the distance and giving commands to his dogs.

The buggy crew, it turned out, had got stuck on the rain-bogged track and had to haul each other out in turn. I looked over at my chestnut horse, tied to the fence, catching some shut-eye before she was needed again. She didn't get stuck in the bogs, I thought. She instinctively knew the right spot to cross, could be relied on to get me where I needed to go. She had also been my eyes that morning, a source of comfort in the back blocks, and I already felt a bond with her. Give me a horse any day, I thought.

*

As we draughted the lambs off the ewes, the baaing reached fever pitch. Meanwhile, the docking and castrating gear was quickly unloaded – conveyor stand, gas-fuelled docking iron, vaccinating gun, antiseptic spray, testicle rings. I was passed the ear-markers, and lambs were flying down the conveyor before I even knew it, the picking-up team putting their backs into their job and jolting me into action, too. I was tasked with making a small cut on the left ear for girls and the right ear for boys, so that the shepherds would be able to make out their sex when they were draughted later on, and I figured out my technique as I went.

The team found their rhythm fast, and it felt good to be a part of the mahi. The shepherds kept the lambs coming, filling the pens over and over, until the yards were empty and the paddocks were full. The lambs' tails gathered around us. We counted them up, and discovered we'd docked just over 700 in a few hours. And you could tell: we were blood-spattered and covered in mud. A blister was already forming on my thumb, and my lips were wind-chapped, but I was exhilarated. There's nothing quite like that combination of extreme tiredness and intense satisfaction you get from a hard job done well.

*

My work was far from over for the day, though. After scoffing a corned-beef and chutney sandwich, I was back in the saddle, racing alongside Tom in hot pursuit of Sheildsy. Tom and I had a hard time matching his pace, and had to pull out a few canters to keep him in our sights. Our horses responded eagerly to the speed.

When he finally came to a halt at the top of a deep gully, Sheildsy chucked me a sheep rattle, a handy little noise-maker made up of clanging steel discs. 'You take one side of the valley, Carly,' he said. 'Go push the sheep up. And, Tom, you take the middle. I'll be out to the right.' And he was off again, and within minutes little more than a silhouette on the horizon.

'All good?' Tom asked me. He'd first come to the station on the Growing Future Farmers scheme – the same one as young Brossy at Muller – and had been good enough, and tenacious enough, that he'd been offered a job. It was a big station for a young shepherd, and he was therefore a bit more sympathetic to me and my lack of knowledge of the land. Unlike me, though, Tom had a real and deep connection to this farm; his grandmother had previously owned it for many years, and his family ties went back a long way.

And I could see that he was anxious to push on, not wanting to be the last one to the yards, so I replied quickly: 'Yep, sweet. Thanks, Tom.'

Big stations are a real testing ground for shepherds like Tom, who are just cutting their teeth. It takes grit and humility, and it's a bit of a rite of passage to feel unsure of yourself, like you're on the bottom rung. Which was precisely how I felt as I headed off, with no real idea of the location of the next set of yards and my chestnut horse side-stepping every time the sheep rattle clanged.

I didn't get far before I heard Tom, in the far distance, yelling something at me. He was gesturing, telling me to look up, so I did. And there, looking down at me from a bluff right above my head, was a ewe and lamb, as still as statues. I swear they both had 'fooled you' expressions on their faces.

'Move on, please,' I muttered at them. 'You're making me look bad.'

Thankfully, they did as instructed. I breathed a sigh of relief, and continued pushing onwards and upwards, gathering sheep and momentum, shaking my rattle the whole way, until I met up with Tom once more.

'You go gather up any stragglers off to the left,' he said. 'I'll keep this lot moving.'

I gave him a thumbs-up, then pushed on up the hill. But what I saw on the other side made me stop dead in my tracks. There were so many ewes and lambs over there. Did Tom really think I could muster up all of them on my own? For one stomach-dropping moment, I felt utterly out of my depth, completely at a loss as to what to do next. When a muster is fast-paced, like this one was, shepherds are often left to follow a trail of instructions that are eked out on the move. You have to piece things together as you go, like a dot-to-dot, and in the momentum it's easy to make a rookie mistake. But you don't want to be that person.

Stop, I told myself. Think.

And, slowly, my tired brain turned a cog and I worked it out. 'What an idiot,' I muttered. Those sheep I could see scattered wide into the distance? They weren't in need of mustering, because they were the ones we had just mustered up the hill. The paddock was huge, and Tom was pushing the sheep over to the vast green space that I was looking down on.

I'd just learnt one lesson every shepherd seems to learn early on: never think you know what you're doing until you've done it many times over.

*

After meeting back up with Sheildsy, we once again got the sheep into the yards and repeated the draughting-and-docking process. Lambs were flying through the conveyor to an eclectic soundtrack of hip-hop beats, nineties tunes and the occasional rock belter that got everyone singing along. Spirits were high, and as the pile of lambs' tails grew higher, everyone got a bit more grime-covered and a bit more

Young helpers ham it up at docking time.

Down hills are a time for hopping off your horse to give it a break and for stretching your legs.

feral. There was a fair amount of good-natured jostling and competitive elbowing.

'I'm picking up more than you, bro,' Thorston threw at TK, the station cook's teenage moko.

'No way, bro,' countered TK, flashing a wide grin. 'Your sister is faster than you.'

Anika just rolled her eyes and worked on, determinedly picking up lambs twice as fast as the boys, while Lindy chuckled.

We had a few new young ones in the team that day. Zara lived down the road from Mangaheia, and was a keen horse rider. Patrick sported an impressive mullet, and although he couldn't see over the pen he got cracking picking up a lamb and getting it to us with a grunt of effort. Then there was Zara's mate, Liam, aka Tarzan.

'Tarzan's a townie,' Zara announced when she introduced him to me.

'But I want to be a shepherd!' he quickly added.

Tarzan, as it turned out, knew the lyrics to a lot of the songs blasting, and he even had a few dance moves, too. While he took on the task of injecting the lambs with vaccine, he told me more about his master plan to become a shepherd. 'I don't really like school, so I want to leave when

I can and come and work for Leo. I'll hopefully do all the dockings and get really good so he will want me.'

'What about the horses? Will you be into that?' I asked.

'Oh yeah!' said Tarzan, cheeks flushed with enthusiasm. 'I'll do it all. I'll get my own saddle and everything.'

*

When we were done, we found a spot out of the cold wind to drink strong tea in tin mugs and inhale some biscuits and muffins while comparing blisters and bruises. TK told me he too wanted to become a shepherd when he left school. He had horses at home, 'wild ones', he said, that he would jump on and break in. Thorston was listening in. 'That's mean, bro,' he said, full of respect, and TK tilted his chin.

Horses still ran semi-free up the remote river valleys and deep bush of the East Coast. The bloodlines of these 'Māori' or nāti horses were some of the oldest in the country; they could be traced back to some of the very first horses to set foot on the whenua. Ownership of these horses existed on the coast, firmly established by locals but not always obvious to outsiders. The breeding of them, too, could appear haphazard, but in fact if you asked a local they would usually be able to tell you the pedigree, plus a few yarns about the nature of the dam and sire. Hōiho were family up these ways. Time-honoured animals that were interwoven with Ngāti Porou culture.

The ancestral whenua that so many of these East Coast locals rode with such ease and confidence has deep whakapapa. A large pā site, Ratau, forms part of Mangaheia's landscape, on a high ridgeline that runs east to west above Ūawa/Tolaga Bay. The pā's name means 'last hill to catch the sun at the end of the day', and it was once occupied by the iwi known as Tarewa, before they were overthrown by neighbouring iwi, Hingaroa. It's thought that about 167 Māori lived at Ratau, and the pā is just one piece of the much wider cultural landscape of Ūawa.

The early Māori population of Tolaga Bay was large at around 1200, and the main iwi was Te Aitanga-a-Hauiti. With kai resources from the sea, rivers, inland forests and horticulture practices, Tolaga Bay was a place of abundance. At one time, the whānau of Huhana Tamati, Te Kooti Arikirangi's wife, lived at Mangakuku, just above the Mangaheia Station woolshed.

European settlement didn't begin in the area until 1875, when a 7334-acre Takapau block was bought from

18 Māori owners by Robert Noble. A farmer and horse-breeder, Robert paid £1436 pounds for the land (around $250,000 in today's terms), which was considerably more than the Crown was paying for similar blocks in the area at the time. Then, in 1886, Richard Reynolds bought a large portion of the Mangaheia block from the New Zealand Native Land Settlement Company and the Bank of New Zealand Estates Company. At that time, the station was 7200 acres and Richard's sons, George and Ralph, were running about 10,000 sheep on it. George took over the property from his father in 1921 when some of the land, sought-after for its rich silt flats, was subdivided and sold. In the decades following, there was a succession of owners and managers at Mangaheia, culminating most recently in Leo.

Here at Mangaheia, there was a real sense of those who had come before – and, rather than a knock-up-against clash between Māori and Pākehā, it felt more like a blending of two creeks.

It was Al, that station cook's husband, who told me the Māori name – 'the proper name' – for Tolaga Bay is Ūawa Nui a Ruamatua, and Anika helped me unravel the reo. She said Ūawa is an old word, as old as these hills, and as I rode them I repeated the name to myself, pronouncing the syllables in time with my chestnut horse's footfalls, trying to memorise it.

Ūawa. *The Tolaga river.*
Nui. *Plentiful.*
Rua. *Two.*
Matua. *Parent.*

*

The stars were out the next morning, which only made the early start seem more surreal. It did mean I could see a little more, which was certainly appreciated, and I reminded myself as I keenly followed the shepherds up yet another hill to notice where I was, to take it all in. Because this was a pretty extraordinary way to start a day: riding a strong horse up a huge hill in the starlit dark with a crescent moon for company, carrying just a coat, a bottle of water, the sleepiness of the early start countered by the buzz of adrenaline.

We made our way up to the station's highest points, and as we followed a jagged ridge the sun finally came up, bathing the world in light.

'We need to get across that gully over there,' Leo said to me. 'And the best way is to skip over into the neighbour's farm.'

That was Murphy land. I already knew this because I'd been told that the Murphys, as well as being the original European settlers there, were also the ones said to have started the Gisborne-bred horse. All the Mangaheia horses were Gisborne-bred, meaning they followed an original bloodline with its roots in the land I was now riding over. There was a notable contrast between this farm and Mangaheia; it was much more rolling than the craggy upper reaches we'd just come from.

'Mangaheia is the kind of land I like,' Leo said when I commented on the difference. 'Big hills where you can really work your horses and dogs.' He had called Mangaheia home for just over ten years, and had developed a deep love of the East Coast. 'We work hard and fast out here, and I like that.'

Having safely traversed the gully, we headed back on to Mangaheia land, and made our way to a ridge that looked out over Ūawa/Tolaga Bay to the ocean beyond. The maunga, Tītīrangi, sheltered the bay, and out even further was Pourewa Island. From up here, beyond the hectares

Homeward bound and smiles all round knowing that a cold beer is only a few hills away.

The East Coast has the perfect hills for using horses.

of descending hills, I could see Te Upoko o te Ika (Cook's Cove) where, in 1769, Captain James Cook landed aboard the HMS *Endeavour*. And down below us in the flats, where a silver cord of a creek snaked itself to the Mangaheia River, lay the yards.

The view was huge, and the chestnut horse and I took it all in while Leo set his dogs to work. It was quite something, and it was clear he'd put time into his dogs. They were highly disciplined and fit, just like the horses. His heading dog, after one whistle, was ripping off down the hill, the sheep in her sights and her job firmly in her zippy little brain. The huntaway gave it some heft with a resounding bark, and within minutes the ewes and lambs were starting to stir.

'Seems like the perfect way of doing things,' I said to Leo. 'Standing on a hill while your dogs do the work for you.'

He laughed. 'That's exactly right.'

*

With sheep, I was learning, the job was often completed faster in the long run if the animals weren't rushed. As we pushed our mob on, it eventually mingled with the other shepherds', and I could see the morning's work coming together. Shepherds and dogs were spread on the hills, while the sheep quietly filed in from both sides, everyone heading towards the last hurdle of the creek, before getting the mob safely in the yards.

The day rolled on, just as the hills that surrounded us marched towards the sea. We headed to another yard full of ewes and lambs, and then another. And the day's end came fast, as did sleep, and then the following morning too when – you guessed it – the whole process restarted, and we did it all over again. It was exhausting, exhilarating and hugely satisfying work, and it carried on this way for a whole week.

When the last day of that long week came, Leo tallied the lambs' tails in his phone.

'How many?' I asked him, hauling a wool sack full of tails destined for that night's barbecue on to the quad trailer.

'About ten thousand,' said Leo.

'No, really,' I said, sure he was pulling my leg. 'How many?'

'Ten thousand,' he replied. 'Really.'

That meant I'd marked ten thousand ears! No wonder my ear-marking hand was in a permanent claw and my blisters sported blisters.

My respect for this crew had grown and grown in my time here with them. They were tough and hard-working and fun, and so were their horses. Everywhere I'd mustered so far had used horses, but here at Mangaheia it was next level. The only place a quad bike really had on this station was as a way to cart out the docking gear. The Mangaheia horses were the most athletic I had come across yet, and they worked most days, only getting a day off when the shepherds did too. They traversed long and hard terrain, and were absolute athletes, all muscle and hard-working grunt. My chestnut horse had never tired underneath me. She had carried me a massive number of miles over dauntingly rough terrain and hadn't hesitated or given up once. Everyone, and every horse, here was essential. The horses in particular had always had a place at Mangaheia, having served these hills long before fence-lines were put in.

And, even though no one said it, we were all feeling our muscles by the time our last smoko together rolled round. We took a bit longer over that one, knowing our day was almost done, and afterwards flopped down in the shelter of an elbow in the hill, the land and sea our vista, a world away from anywhere.

'I can see why people come to this place and never want to leave,' I said, flat on my back and gazing at slow-moving clouds.

'That's it,' said Al. 'Why would you want to?'

8
SURVIVAL OF THE FITTEST

*Ngahiwi Station, Tairāwhiti/Gisborne,
and Awapapa Station, Tiniroto*

'Shall we call in on Bruce?' said Lindy as we drove away from Mangaheia.

'Be rude not to,' I replied, smiling. I'd been hoping she might suggest that.

Bruce Holden lived on the outskirts of Gisborne, and was renowned for breeding Ngahiwi horses, which were famous in the horse world and an eyeful of beautiful — proud, flashy and as majestic as they come. 'Bruce is an absolute legend,' Lindy had told me earlier. 'He'll be able to tell you the origin story of those Gisborne-bred horses.'

When we arrived, Bruce bundled his old friend Lindy into a tight hug, then turned to me. 'I reckon lunch can

This way leads to horses. Ava Thomas

wait,' he said with a sparkle in his eye. 'Let's get on a horse.' I could see right from the get-go that there was no moss on this 80-something. Bruce Holden was a hoot.

While Lindy disappeared inside with Bruce's partner, Faye, he and I went and started saddling up. 'How was Mangaheia?' he asked. 'Are the horses any good still?'

'It was great,' I replied. 'And yes, the horses were great too.'

'Good,' he said, reaching down to tighten his girth. 'That station has a place in my heart. When I worked there, I was just a young buck, and I wanted to breed my mare with the stallion. But the boss said no. So, when he was in town one day, my mare "accidentally" got in with the stallion and, boy, did I get in trouble!'

I laughed as I mounted Captain, a tall and leggy bay whose energy I immediately felt rising up through the saddle.

'But it was worth it.' Bruce winked at me.

Then we were off, down the gravel track leading to more of those East Coast hills. As well as being windswept and rugged, these ones were slip-strewn and rain-battered. The weather had thrown all it had at the region, swelling rivers and bringing down whole hillsides, and the going was hoof-sucking muddy, but our horses negotiated it all with a high-stepping stride and a keen jump that cleared ditches and creeks.

For once, we weren't chasing any stock, and that was a real luxury. Bruce had hung up his mustering hat when he leased out his land, Ngahiwi Station, in 2014, but he had years of pushing cattle and mobbing up sheep under his belt. And he never did quite hang up his saddle. He flew along on his horse with a long-legged, straight-up-and-down posture, his hat magically staying put, exuding a confidence and ease that defied both his age and the gnarly weather. Rather than shifting onto a quieter, smaller horse in his later years, Bruce had stuck with what he loved: an athletic horse with a spirit that equalled his own.

We paused at the top of a climb, and he extended an arm to capture the sweep of the view. 'There's the city, there's the coast and here we are,' he said.

'In the hills,' I said.

'In the hills,' he agreed. 'Where nothing else matters.'

*

The Holden family had a long history in both farming and the East Coast, two things that went hand in hand with horses. 'Horses are in my blood,' Bruce told me. 'I just love them. I always have. It's an instinct, I suppose.' His forebears had started out at Tikokino in the Hawke's Bay, not far from Smedley Station, back in about 1859, then Bruce's grandfather had come up here to Tairāwhiti in about 1880. In those days, there hadn't been any road access to the station – there was just a track over the hills that had to be traversed on horseback. Bruce's grandparents were both accomplished riders, and his grandmother once accompanied her husband on a trip from Wairoa to the Bay of Plenty, all on horseback and with the mighty hills of the Urewera to get through. That grandfather died in 1941, Bruce told me, 'and I popped out in 1943.'

To begin with, the station had been 7000 acres but, as with many family stations, the land had been divided up and sold over the years. 'So now I'm the only Holden here,' Bruce said, 'and we have about 1500 acres.' He held onto enough land for his horses, and the paddocks were full of them – mares and foals, three-year-olds waiting to be broken in, others waiting to be sold. Ngahiwi horses could be found in all sorts of arenas – mustering the hills, hunting, and showjumping and eventing at the absolute top level. They were known for their athleticism, their brains and for having what Bruce called 'a bit of heat'.

'I don't like a lazy horse,' he told me. 'They need to have a bit of something to set them apart. The breeding is important, but when it comes to a horse, a lot of the time I rely on my gut feeling.'

Bruce didn't mollycoddle his horses. They were put onto the hills right from the start, whether they were destined to work stock or the show ring. 'I have the philosophy that it's the survival of the fittest,' he explained. 'So that they become physically strong, but they also have the intelligence of survival. We don't pamper them here. The mares come in to foal, and then they go out. We only have eight acres of flats here, and then all the rest is hills.'

Then he turned to me, clearly sizing me up. 'I reckon you know what you're doing,' he said with a mischievous grin. 'Shall we go up a gear?'

I didn't have time to respond before we were off again, flying uphill, sliding down the other side and sloshing through the boggy base. The going was up and down in equal measure, and flat land was nowhere to be seen.

By the time we ambled back into the yards, the heat had gone out of the horses. 'It's good for them to get a bit of a sweat up,' Bruce told me.

And I was impressed at how quickly Captain settled from high speed to a steady, reins-at-the-buckle walk – but I could tell he still had plenty in the tank.

*

Over lunch and a cuppa, Bruce shared with me all he knew of the famed Gisborne horses. The story started with Edward Murphy, an Australian horseman who had

MANGAHEIA

Mangaheia River winds through the station.

MANGAHEIA

Mangaheia Station stretches out to the east coast.

Dogs are an essential part of the Mangaheia Station team and manager Leo is a champion dog handler.

MANGAHEIA

Smoko time at Mangaheia Station is about flopping in the grass for some well-deserved rest and kai.

Leo gives his horse a break and his legs a stretch as he heads down to where his dogs have a mob of sheep contained.

NGAHIWI

Bruce Holden in his happy place surveying his herd and talking about bloodlines.
Ava Thomas

Ngahiwi horses are sturdy, strong and agile. Ava Thomas

NGAHIWI

A mob of Ngahiwi broodmares with foals at foot. Ava Thomas

Ngahiwi mares and foals turn tail. Ava Thomas

MIDDLEHURST

Willie and his team of dogs in the back reaches of Middlehurst Station. Francine Boer

Smoko time reflections on the day, tired dogs and the most wonderful view.

MIDDLEHURST

Susan and Willie take a moment to survey their land. Francine Boer

The light from Old Middlehurst Hut is a welcome sight after a long day in the saddle.
Francine Boer

MIDDLEHURST

Susan moves cattle along the side of a hill, sending her heading dogs out to keep the mob tight. Francine Boer

Jonty catches a string of horses in the early morning light. Francine Boer

MIDDLEHURST

Autumn musters can mean frosty, bone chilling starts to a working day. Francine Boer

ST JAMES

An old stone cottage out at Okuku Pass.

ST JAMES

The St James sale in full swing. Ava Thomas

Backcountry huts are a home away from home for musterers.

ST JAMES

Lyndon Morris has a long association and love of the St James horse.

Lyndon Morris in storytelling mode as he starts in on another mustering tale.

ST JAMES

A couple of young St James horses tuck into some good pasture.

BREAKING IN

Getting a young horse used to the saddle before getting on. Ava Thomas

BLUFF

Pitt Island from the air. A big rock in the middle of the ocean.

Giving the pilot a hand to push the Pitt Island plane out onto the tarmac.

BLUFF

Paige Lanauze and her beloved foal Moose.

James Moffett, author Carly Thomas, Reuben Moffett and James (Moff) Moffett after a day's ride.

BLUFF

Everyone gets involved at the Bluff Station muster and all the kids come along for the ride too.

The long march to the Bluff Station yards.

BLUFF

DD on his mare Concrete Lady who has a foal at foot.

James and Annette on two Bluff Station bred horses with Jame's bearded collie dogs in tow.

lived at Manutūkē, just south of Gisborne, in 1884. He'd wanted the best type of horse to carry him between the several hill stations he owned. For Edward, that meant a horse that could provide comfort, speed and stamina. He was looking for something other than the heavy-set Clydesdales that were the norm at the time. And it sounds as though he was mighty intrepid – though Bruce put it a slightly different way. 'Eccentric is what I think you are when you're passionate about something,' he said. 'When it's not about money, when it's about your love of it.' Whatever you want to call it, that particular thing was in Bruce too. In the twinkle in the eye when he talked about horses.

In 1902, Edward ended up in Kentucky in the United States, and there he got himself an American saddle horse. The five-gaited stallion, named Kingston, was the first of its kind in New Zealand and had been bred to have a very particular walk – it covered the ground in a vigorous, active stride that the rider could just sit to comfortably for many miles. The horse's strong hoofs were also a handy thing in Tairāwhiti, where the weather meant the going could be soft one day and hard the next. And the stallion was a good-looker, too, much flashier than the station hacks in New

Zealand paddocks at the time. I can only imagine what a stir Kingston's arrival in Gisborne must have caused.

Back when Bruce had first heard this part of Edward's story, he'd been so intrigued he'd taken himself over to Kentucky to find out more about this purchase. That was in 2004. But, when he got in touch with the American Saddlebred Museum, the woman he spoke to told him, 'I very much doubt that happened, Mr Holden.' He asked her to look a little further – he was that adamant about the bloodlines of Kingston, with his strong shoulders and high carriage. So she did … and soon rang back, apologising and confirming that yes indeed, Edward *had* exported Kingston to New Zealand. Furthermore, he'd been the first person to ever export an American saddle horse from the United States. And, she said, the stallion had come all the way by boat, accompanied by a groom, to the wild East Coast.

Kingston covered the ground and was strong up front, but Edward wasn't done. Next, he needed a horse with a strong back end. 'A big butt,' as Bruce put it. 'So off he went to Vienna, and he stood in front of the gates of the Emperor of Austria's Spanish riding school.' A Lipizzaner is what Edward had set his sights on – a horse that was broad and powerful 'at the arse end', as Bruce put it. But Edward

Bruce Holden is his happy place, amongst his horses.

A good eye on a horse is what Bruce Holden looks for. Ava Thomas

was informed that the stables absolutely did not sell colts. So what did Edward do? 'Well, the story goes,' Bruce told me, 'that Murphy camped outside the gates for about two weeks. He said he wouldn't go until they sold him a horse.' And, eventually, they did. That stallion was Maestoso, a magnificent white horse who became the first Lipizzaner to ever high-step onto New Zealand dirt.

That was the beginning of the story. From there, Murphy's horses started to get a name for themselves. 'They were tough and they could go the distance,' explained Bruce. 'The races were very important in those days, and one day Murphy and his mates rode all the way to Wairoa for a race meeting. His horses were very gutsy, and I think that's what the Gisborne horses are still known for.'

*

In the eighties, Bruce noticed those old Gisborne horse bloodlines were in decline. That's when he started seriously breeding horses himself. 'I decided to resurrect them,' he told me. He wanted to continue those bloodlines. So he got himself a foundation stallion, a great big grey named Panikau (after the Murphy land it was born on) from

Edward's grandson Peter Murphy. It was Panikau who sired many of Bruce's early Ngahiwi horses. They were bold and strong, but also had brains. 'They are horses that you can't take for granted,' he told me. 'If you kick them in the guts, then they might put you on the ground. They're not for beginners. The headpiece is number one – the brains. A good brain and a good eye. A good, open eye.'

It was the science and also the magic of it all that really got Bruce hooked on the genetics side of things. In 2003, he began having his mares artificially inseminated, but the success rate initially wasn't great. 'Then I met Doctor Lee Morris,' he said. 'That changed everything.' An expert in horse fertility, Lee took Bruce's in-foal rate up to 100 per cent. 'She's one of the best in the world,' Bruce told me, then his eyes flashed. 'And she's another one who is in it for the love of it. The people who are brilliant at things have to love it, don't they? We are all a bit eccentric.'

He laughed, and so did I, because there's no escaping it: yes, people who really love horses can be a little bit in-a-good-way mad.

*

The Last Muster

After waving goodbye to Bruce and Faye, Lindy and I got on the road once more. This time, Lindy was dropping me off at the Campbells', in the hills of Tiniroto. Our route followed the wide sweep of the Wairoa River, and I had to jump out to move a few road cones before we reached a part of the road where a sign announced it was closed. Alex Campbell had already warned me about this on the phone, and given me instructions for getting through. Cyclone Gabrielle had hit hard here earlier in the year, and the now-placid Wairoa River had just months before burst its banks, devouring bridges and demolishing already fragile road networks. We bumped along, dodging yawning potholes, before finally turning into Awapapa's gate.

I farewelled Lindy, then hopped out of the car. It was Alex's 96-year-old mum, Isabel, who greeted me when I stepped into the farm kitchen. 'I'm Grandma,' she told me, with a beautiful smile. Isabel came to Awapapa Station with her husband, Jim, in 1967 and after decades of calling the station home, she still loved being among the daily hustle and bustle – and she still contributed. Isabel had two empty lamb bottles tucked under one arm, and a tea towel over the other.

'And I'm Alex,' said the tall man with a wild mop of white hair who'd just walked into the kitchen with a little fox terrier at his heels.

Chairs were pulled up, cake was produced and we spent the next hour talking horses. Alex got right down to it, starting at the start, on the day in 1967 when his late father, Jim, had brought his family from Southland up to this isolated farm on the East Coast. 'It was after the war. Dad was 40, and he had got together a deposit to secure Awapapa Station through a Lands and Survey ballot settlement,' Alex explained. 'We got here, and there was no equipment, nothing except for one horse. That was the total sum of it.' Alex was six years old, and that was his first encounter with a horse. 'Something started off in my heart,' he said. 'It fitted, and it went right through my life.'

Alex had known the late Paul Johnson, a man who dedicated his life to breeding a good line of East Coast horses. 'The Paul Johnson horse practically became a category of its own,' Alex told me, arms crossed tight over his chest as he leant back on his chair. 'People have always been proud to say they have a Paul Johnson horse.'

It was a bit of equine alchemy that Paul conjured over the decades, and his particular eye for a good horse played a part, too. 'He came out with something all his,' Alex told me. 'And he said to me once that, as long as the horse has a good head, "the body will follow". He was a great horseman and, while he was a hard man, he loved his horses more than anything. And he sure could breed them.'

In his heyday, Paul had about 500 horses at one time on his vast land in Ruatoria, and sold off his unhandled stock in annual auctions. His horses were sought after for mustering, known for their toughness and steady nature; and, like Bruce's Ngahiwi horses, they proved themselves worthy on the hunt field too.

When Paul was still alive, Alex had sat down with him and discussed how he could help in keeping 'the dream alive'. 'I came away with a good colt and ten mares, and that was the foundation to get things going and to honour Paul's work,' Alex told me. 'Those bloodlines are just so important, and I don't want to see them diminish.'

In those bloodlines was Cleveland Bay, a breed that originated from the north of England and was one of the oldest in Britain, having been first established in the 1700s. The Cleveland Bay was listed on the world rare and endangered breed list, so to safeguard the future of Paul's original breeding, Alex imported mares and a colt from

Australia. Additionally, Paul had also added thoroughbred lines into his heavier herds, and he prized mares from nearby Mangaheia for their quality.

'And now I'm breeding these wonderful horses,' said Alex. 'We call them Hangaroa Land Cruisers.'

*

Hangaroa Land Cruisers, Alex told me as I sat in his ute on the way to see his foundation mares and a young mob of foals, 'have put some hope back into our lives'. During the Global Financial Crisis in 2008, he and his wife, Megan, lost their farm in a mortgagee sale; now, they manage the farm and still live there, but they have had to rethink their entire life plan. 'We lost everything that not only we built, but that my father did as well,' Alex said. 'But we started from nothing, and we have an open mind. My father taught me that – an open mind, and a bit of out-of-the-box thinking. Those were my inheritance. It hasn't been an exercise for the faint-hearted, but we just haven't given up.'

As we came to a stop in the middle of the paddock, the youngsters gathered around the ute. Long-legged and cheeky, they licked the wing mirrors, rubbed themselves against the bumpers and pushed their whiskered muzzles through the windows.

'How can you be down when you have this?' Alex said. 'They lift you up, don't they?'

I had to nod at that. There were 12 or so yearlings surrounding us, and they were delightful, nosy, mischievous – and an important part of Aotearoa's equine history. 'What we are looking at here is very special,' said Alex. 'These are true East Coast stationbreds. What that means is that in these bloodlines are those original characteristics. These horses have been bred to withstand the environments of sheep and cattle stations. They are a fact of life and a necessity up this way, and their strength, reliability, stamina, honesty, intelligence is what the stockmen relied on. They trusted their lives to these horses, and that's what we are trying to save.'

A big mob of mares, all pregnant and as round as barrels, were dotted throughout the large paddock; Alex proceeded to name each and every one, and map out an intricate lacing of genealogy. Just like Bruce, he was passionate about drawing lines under the ancestral paths these mares had taken and where they were leading. The mares before us were mostly big-boned and noble-headed, with those Cleveland Bay bloodlines really showing through.

Bruce's mares and foals live as a herd in the hills behind his house. Ava Thomas

Running as a herd in the hills when they are young makes the foals strong and hardy. Ava Thomas

'The preservation of these horses is diminishing, and these stationbreds are right from the heart of Ruatoria,' Alex said. 'I'm very lucky to have them here, and I hope that I do Paul proud.'

Then he paused for a moment, taking it all in. 'It's exciting, isn't it?' he finally said. 'All the possibilities in this paddock.'

*

After heading back through the prancing and dancing youngsters, we made our way once more along the potholed road. The scars left by the cyclone were still raw: slips gashed the hillsides, and the riverbank had been stripped back to its bare bones. It had not been an easy existence for anyone in this part of Aotearoa, and for the Campbells the cyclone was just one more thing they'd had to brace themselves through. 'The horses have been a way for us to do something that's for our future,' Alex explained as he drove. 'Something away from all the negative stuff. Horses can do that for people, I think. They can give us something to rely on and trust.'

When we arrived back at the house, I was introduced to Chloe, a young horsewoman and keen shepherd who Alex and Megan had taken under their wing. 'Did you get the big horse tour?' she asked.

'I sure did,' I replied.

Chloe was, she told me, one of the Hangaroa horses' biggest fans, and she played a main part in their day-to-day training. Once started, the horses were put to work mustering on the hills of Awapapa, and Chloe had also been taking them to the odd show and hunt to give them some exposure to that kind of life as well. Sam Sidney, a legendary horseman and handler, was also part of the Hangaroa team, and he gave the young horses their first education, assisted by young East Coast horsemen Tama, Blaeq and Tyrone. Yes, Tyrone – the one and the same I'd been so impressed by when I saw him breaking in his youngster, Āio. I hadn't been at all surprised when I'd found out he had won Smedley's horsemanship award on his graduation, and he'd since taken a job at Taimoti Station, which neighboured Awapapa, and worked with Alex's horses in his time off.

'I'd love to see Tyrone working on a Hangaroa horse,' I told Chloe and Alex.

'I'm sure we can arrange that,' Alex replied. 'I'll get him in the yard with Kenworth. He's what Paul would call a freak, that horse, but in a good way. He's an out-of-the-box horse.'

*

When we arrived at the yard, Kenworth was standing there like it was a stage set just for him. Leggy, broad and with that unmistakable Paul Johnson head – wide, wise and with a kind eye – the dark bay colt strutted around like he was the king of Tiniroto.

'You're still chasing those horses around, eh, Tyrone?' I said when I saw him.

'I am!' He chuckled as he climbed over the railing and picked up his stock whip.

Tyrone already had a few Hangaroa horses down at Taimoti, and was giving them some mileage in the hills for Alex. 'I get to use horses nearly every day where I am,' he told me. 'We need them, with the state of the tracks after the cyclone. We can go anywhere on them – over and under. And then I come out here and work with Alex's horses, so it's pretty good. I'm getting to do what I love on some great horses.'

Now, Tyrone turned his attention away from chat and towards working with Kenworth. He started moving the big colt forward, using a dip in his shoulder to switch direction and a face-on stance to get Kenworth to stop. It was a dance, and it took Tyrone about ten minutes to get the colt to the golden spot of 'facing up' – in other words, turned directly towards Tyrone, chest squared, eyes on Tyrone as his safe place and centre of communication.

Tyrone took a pause by the railing. 'The Hangaroa horses are real quiet and good going,' he said, not taking his eyes off Kenworth. 'They don't take long to connect up and face up.'

Next, he fashioned the whip into a makeshift halter, slipping the cord over the colt's nose. Kenworth initially backed up, but soon bent this way and that when Tyrone asked him to. Getting a hand on the horse was the goal, and slowly, slowly, softly, softly Tyrone touched the horse's shoulder, then his neck and eventually his place of vulnerability: his big, open face. 'There we go, boy. That's it,' Tyrone murmured, stroking the big horse's wide white blaze.

The real halter went on next, then Tyrone cycled through cracking a whip right beside Kenworth and rubbing a flag all over the colt's body, all the while repeatedly inviting him back to the centre – 'the bubble', as Alex called it. Finally, Tyrone quietly put his weight onto the horse's back, and Kenworth held his ground, bending his long neck to check out what was going on. Tyrone turned him to the left and to the right, then jumped back to the ground, and patted the horse. Kenworth bowed his head, and Tyrone turned and grinned at me. 'Pretty neat, eh?'

It really was. In the 20 minutes I'd been there, holding my breath and clinging to the rails, I'd witnessed something special. Kenworth had gone from a snorting, head-held-high, tense-muscled beast to a horse that was bending in, listening and stepping towards Tyrone. It was that magic once again. A young man and a young colt moving forwards, together, in a world where, in the right hands, a horse held possibilities. A world that could open to both of them.

And right then, right here, that's all that was needed. In a dusty yard on a back-country road, a legacy was being continued in the most humble of ways and in true East Coast style. It was a starting point, a continuation and a looking-back all jumbled into one, and it was the time-honoured art of turning out a good horse that this wild part of Aotearoa was known for.

9
ANYTHING IS POSSIBLE

Middlehurst Station, Awatere Valley, Marlborough

My mustering adventure had turned a full circle of seasons, and a year on I found myself back in the South Island. This time, I was on the western flanks of the mighty Inland Kaikōura Range, where the hills bowed down at the feet of gloaming mountains, where the skies were just that bit wider, and the ups and downs that much further.

Autumn was Willie and Susan Macdonald's favourite time of the year, a chance to get out to the back reaches of their high-country station. And, although the station had been theirs for more than two decades, Susan still couldn't quite believe it. Middlehurst consisted of 16,550 hectares, with the top reaches hitting 2400 metres above sea level, and as Susan told me when I first arrived, 'We never thought in our wildest dreams this could be ours. And now we are living that dream.'

When the couple purchased the station in 1998, the first big job facing them had been the autumn muster – and they spent it in open-jawed wonder at the huge country they'd just acquired. They had taken over right at the time of a dragging drought, when both rabbits and broom had taken hold, but even with all the hard work starkly laid out before them, Susan said 'it was just so exciting'.

'On that first muster you must have been 29 years old?' Willie said, seeking and receiving a nod of confirmation from his wife. Like Susan, he was tall and athletic, with a healthy glow that told of days spent outdoors. 'Yeah, we took a huge risk,' Willie went on. 'A risk we didn't really see at that time, because we were young. But now, looking back, we can see it, and it was big.'

'It sure was,' agreed Susan. 'But at that time, every corner we went around, every ridge we looked over, there was just so much good country. No scrub, just open country. And so, so much of it.'

In their time at the station's helm, the couple had always made protecting the fragile land, with its fertile volcanic soil and wide-ranging weather, a higher priority than having a huge production rate. 'We don't look around and pat ourselves on the back and say we own all this,' Willie told me. 'We look around and say, "We are responsible for all of this."'

*

We set off the next day for the station's annual four-day cattle muster, riding out towards Old Middlehurst Hut under bright sunrays that gave everything a sheen of optimism. Willie and Susan's 17-year-old niece Grace was with us. She was on school holidays, and had come to the station to get away from boarding school and outside of cell-phone reception. And, although she told me she hadn't ridden much at all, I could see she was a total natural. She and Fran, a Queenstown photographer who had also tagged in on the muster, were on huge horses borrowed from the station's farrier and herd scanner, Earle – yes, the very same one who'd told me about the magic of Awapiri's Swale Country. Both horses had come with heavy tooled-leather stock saddles, sheepskins for saddle blankets and hackamore bridles, and the two women looked the real deal – gorgeous cowgirls perched on top of their mighty steeds.

Willie surveys his land for elusive cattle. Francine Boer

Middlehurst Station looks dramatic with incoming rain clouds shrouding the ranges.

I, meanwhile, was on pint-sized Bushy Pony, who I'd borrowed from Upcot Station down the road. Bushy Pony was the mum of Raisin – the pony Bridget had ridden on the Awapiri Station muster – and had a bit of a legendary status up this way. She was feisty and from unknown origins, but suspected to have at least some Kaimanawa blood in her. Kaimanawa horses are New Zealand's last remaining true wild horses, and inhabit the vast Central Plateau of the North Island. Every year, they're mustered in and a portion sold in order to maintain the herds at a sustainable level. Once broken in, Kaimanawas can become brilliant horses, with clever brains and big personalities. Bushy Pony's real name was actually Heidi, but she got her nickname for acting all 'bush pony'. She was gorgeous – a bright chestnut with a flaxen mane and tail just like Raisin – and behaved like she was four times her size.

'Are you getting withdrawals from your phone yet, Grace?' Willie asked cheekily not long after we got going.

'*Really*, Uncle Willie?' Grace responded, with the most wonderful eye-roll.

Willie grinned his trademark grin. Family was important to both him and Susan. They each lost a sibling when they were young – Willie's brother in a horse accident, and Susan's sister in a car accident. Walking with grief had shaped both their outlooks on life, and Willie pondered, during a quiet moment while I rode alongside him, whether it was something that, in a way, united them. 'We both know the importance of living every day, and doing the things that we want to do,' he said thoughtfully.

As we passed through Ewan's Block, Susan informed me that it was named after Ewan Stevenson, who owned the station in the 1950s in partnership with his brother Jack. In 1995, Jack's son Robbie took on the station, and he then sold it to Willie and Susan. What that meant is that the Stevenson family had farmed this land for over 90 years; and they were still an integral part of the valley with Bill, another of Jack's sons, owning nearby Upcot Station.

Middlehurst was originally named by Sir Frederick Weld when he had passed through the valley in 1850, en route from Lyttelton to Blenheim. He became the land's first European owner, and he waxed lyrical about the basalt peak of Mount Lookout, which stood in the centre of Middlehurst, saying it held 'the richest vegetation I ever saw'. In the late nineteenth century, the land was then taken up by the Mowat family, who eventually owned a number of properties in the Awatere Valley, though Middlehurst was their first. There was no road to the Upper Awatere at that time, and all the station supplies had to be packed up the riverbed with wagons and packhorses. When it came to shearing, the sheep would be mustered down the valley to be relieved of their heavy fleeces. From there, the wool was transported by horse and wagon, and by bullock drays to the Boulder Bank, and most probably the journey's end would have been a seaside tavern at Picton.

The Mowats were typical valley settlers of the time: cultured and educated, with Old World views about the right to occupy land. As T. Lindsay Buick wrote in his 1900 book *Old Marlborough*, 'the liberalising influence of colonial life had not mellowed into even partially accepting the democratic doctrine of "the land for the people". These men held the great valleys in the hollow of their hands.'

Thus, the high country became laden with the names of settlers such as Mowat – blocks, ridges, gullies, bridges, huts and homesteads all bearing layers upon layers of colonial lineage. Meanwhile, beneath it all, lived the pre-colonial Māori names – some buried, and some defiantly remaining, such as the one that rises above the rest, Tapuae-o-Uenuku. The maunga a staunch reminder that names might be ignored, or overwritten, but they survive nonetheless.

*

After we had gone through Melhops Block (named after a fencer), the country switched up a gear, getting steeper as we snaked our way upwards. The horses powered into their work, sweating their way to the top, with each vista giving up another secret. Middlehurst was vast all the way to the base of the Kaikōura ranges, an ocean of land with peaks like paused waves. It felt overwhelmingly endless; it was hard to imagine that towns and cities could possibly exist beyond such an expanse. It was a whole world out there, one where possibilities lived even if people didn't.

Susan was no stranger to remote. She'd been brought up at Halfway Bay Station, on the shores of Whakatipu Waimāori/Lake Wakatipu, where access had been only by boat or air and, in Susan's words, there were 'no other kids anywhere near us'. She was lucky if she went to town – Queenstown – once every six months. And she loved it.

Her parents bought the 47,000-acre station in 1974, and spent the first month making the unlived-in homestead habitable. 'I would have been six years old, and the oldest of six kids,' Susan recalled. 'The house was full of possums, there were holes in the floor and there was only a coal

Smit takes an opportunity to re-fuel. Francine Boer

range.' But it quickly became home. A pony and a milking cow were brought over on the boat, and there were already some mares and a stallion at the station. And one of the first things that horse-mad Susan and her sister did was to clean out the old cobbled stable. 'They certainly were good days, and there were always horses in my best memories.'

Her dad would let them skip correspondence school whenever he needed a hand in the hills, and he rode a horse called Tim. Meanwhile, Susan and her siblings learnt to ride on a pony that, in Susan's words, was 'bloody horrible'. 'She would bolt, put her head down and we would fall off. But we would just hop back on.'

As a teen, Susan was sent to boarding school, but when she came home for the holidays, she says, 'The first thing we would do was go and find the horses. They would run the length of the Lochy River and we would have to bring them in. We were never allowed to take a motorbike. We'd take bridles and ride them bareback home.' Sometimes, she'd have friends in tow, who got biffed on regardless of their riding prowess. 'It was either that or another walk back,' Susan told me. 'The horses would often take off on them, of course.'

Susan and Willie had four kids, and out of necessity they'd all ridden from an early age, too. They learnt pretty much the same way their mum had: 'They got put on a horse, and off we would go. Work is a way of life out here and they were always a part of that.' Initially, the Macdonald kids also did correspondence school, but Susan admitted, 'Teaching was never my thing. Willie would say, "Come give us a hand," and the kids would get chucked in the truck with their books.' Eventually, a schoolhouse was built on the Macdonalds' property and a teacher was brought in; when that happened, the whole area benefitted, as the rest of the valley kids came to the Middlehurst School for their education as well. And Susan was happy to be relieved of her teaching duties. 'I know where I would rather be,' she told me – and she didn't have to spell it out. I could see quite clearly she was most at home in these hills.

As I watched, she fast-paced it on her ridiculously named grey, Tony (when I'd laughed at his name, she'd simply exclaimed, 'Tony the pony!'), down to where a group of cattle huddled at the base of a ridge. These cows and calves had been in this block for about a month, and they looked at us like we were an invading army unsettling their peace.

Meanwhile, I followed Willie along a narrow ridgeline with scrub at its base. He told me he wasn't allowed a pony when he was young. After his brother's accident, his father was understandably wary, but Willie had always wanted one, and when he left school to start work as a shepherd he got his chance. He cut his teeth up the Awatere working for Kit and Margie Sandall at Upton Fells, then headed to Otago. There he met Susan – they were both shepherding in Tarras – and together they went on to manage Cecil Peak and Mount Nicholas, two stations with a long history in horses. 'I wouldn't say I'm the most proficient of riders,' he said, 'but I get the job done, and I enjoy it.'

He dropped down into the gully below, leaving me to continue solo along the ridge, picking up a few straggler cows and making my way up to where Grace and Fran waited on the track, ready to push the cattle along. The sun had stuck with us, warming our backs, and the dry grass shone gold. Bushy Pony and I paused on the side of the hill, and watched Willie working the cattle below. His shouted commands to his dogs echoed against the hard hill faces and barks filled the gullies, as ever so slowly the cattle came together. To begin with, they were impossibly far flung, but gradually and systematically they were worked into a tight mob. Not rushed, but quietly controlled. It was a beautiful dance of persuasion and coercion. At last, we had a decent

The cattle begins to string out in the right direction. Francine Boer

mob of cattle moving in the right direction, and the muster felt like it had really begun.

*

When Susan and Willie had first begun looking to buy a station of their own, their list of requirements was short: it had to be in the high country, they wanted somewhere suitable for merinos, and – most importantly – it had to be isolated. Middlehurst was absolutely that. Even with a decent amount of high country under my belt, Middlehurst was just so relentlessly huge. It was hard to take in the scale, land that never quite allowed you to get too familiar with it.

From where we were now riding, the Inland Kaikōura Range stood right in front of us, and it looked like it had been superimposed on to the sky. We were on the western flanks, above the Tone River, in a block called Sankey's, 8000 hectares of mountain country which had once been part of Muller Station.

'How lucky are we?' said Susan. 'Out here, we haven't had to answer to anyone apart from ourselves. Not many people get to do that.'

It took Susan and Willie two years to find this land. 'We still have an old fruit box filled with all the farm-sale brochures we looked at during our search,' Susan told me.

Willie had grown up in Te Anau, where his family still farmed – and where his grandfather had reportedly bought the pub and sold the licence to stop the musterers from drinking the place dry – but the couple hadn't been that keen to farm down there, because it was flat. Then Susan's brother had mentioned Middlehurst. He'd been on a muster out here, and had heard that it was coming onto the market. So Willie and Susan arranged a visit. 'They were shearing hoggets at the time,' Susan said, 'and we looked into the yards, and they were huge animals. We thought, Well, this must be very, very healthy country.'

It was only with the help of one of Willie's uncles who had no children that they were able to secure Middlehurst for themselves. He ended up putting money in for the couple, but as Susan said, 'We had to convince him, and we didn't show him out here at first. The hills would have freaked him right out.' She laughed, eyes sparkling. 'We were very fortunate. It was amazing that we got it. It felt really amazing.'

*

As we rode, Bushy Pony was forever on the lookout for cows. She would nudge me impatiently with her nose when I wasn't going fast enough for her, and the clever little mare even perked up her ears when Susan's dogs flushed out a string of baby pigs and they dashed off in a comical line of squeals.

The cattle were not proving to be great fans of going upwards. On one big, steep uphill push a few, sick of being coerced, charged head-down at the dogs, breaking away from the mob. It took a bit of manoeuvring on the horses and some swift moves from the dogs to get them back on track and up the steep face. We dug in our toes, and kept pushing them along. Here, things were done with persistence and at a pace which didn't put pressure on the animals. Willie and Susan just weren't the sort of people who got in a flap. 'There's no use,' said Willie. There was an endless supply of daunting hills, after all, and this was their life.

As we pushed on, the going got more and more challenging. The horses took it all in their stride, adjusting instinctively to the terrain, putting their heads down and straining their muscles, utterly focused on finding the way through. I couldn't help noticing how their determined and no-fuss approach mirrored the Macdonalds' own: if a hill had to be climbed, you climbed it. Some way, somehow. Just get on with it and get it done.

It was a quality that had served Willie and Susan well in their 20-plus years at Middlehurst. Acquiring the station was a dream, but working it into a reality had been hard work. 'We arrived here, and it was literally a blank canvas – understocked and undeveloped,' Susan told me. 'But we were young and we had a lot of energy.' The couple didn't have a big team of workers, and in the early days they went it alone. They'd started fencing when they first arrived … and they were still fencing. It was neverending. But this was the life that Susan and Willie had both chosen and wanted. A tough and relentless life spent in constant uphill slog, but also a life full of possibilities as vast as the land itself.

Some people seek out an easy path, but not these two. They were in it for the thrill of the challenge.

*

The last section of our ride towards the hut took us through an extraordinary part of the station, where a strip of wilding tōtara and hebes were lush and thriving. Willie also pointed

Willie at Old Middlehurst Hut after a long day of mustering.

Ben and Smit hunt for cattle in the vast and seemingly never-ending hills.

out some native broom, a good indicator of how well the valley was doing.

He and Susan had put a QEII covenant on this area, he told me, to protect the tōtara forest. 'It runs up this gully, and it's a hell of a challenge keeping the deer out of it. Fences get ripped out in bad weather, so it's an ongoing thing to protect it. But it's something worth protecting, and the bird sound up here is just incredible.'

Both Susan and Willie took their roles as caretakers of this environment seriously, and chose to look to the future rather than get stuck in stale and dated mentalities. For them, climate change was front of mind. 'If we don't have the ability to react,' he told me, 'we might discover one day that we've spent 20 years farming in the wrong direction.'

In the early 2000s, he and Susan had opted out of a tenure review, in which their leasehold land could have been bought back by the government for conservation. The couple felt it didn't fit with the farming model they wanted. 'We want to look after Middlehurst as a whole property,' explained Willie. 'Working with nature, rather than against it. And we feel like the best guardians to do that.'

*

It was cheeky Bushy Pony who gave me a hint that we weren't far from the hut. She communicated to me that she was just too tired to carry on – refusing to budge, nudging my foot in the stirrup as if to say 'Get off', and putting her mood indicators (her ears) rigidly back. She was all drama, so I conceded; I jumped off and walked. Then, the moment the hut came into view and we were on the home straight, she miraculously entered chug mode and determinedly started overtaking the big guns by jig-jogging when she could, all the while tossing that blonde mane impatiently. Oh, now you've got energy to burn! I thought. I'm sure that trickster pony had smelt the hut – and her kai – before I did.

Old Middlehurst Hut was in a little dip of a gully, tucked into an elbow-bend of a hill with Middlehurst Stream feeding through it. Climb that hill in front and you got a grand view down the Awatere River. Beyond that, Mount Upcot loomed, and at our backs was Mount Winterton. The hut was old, and the question 'How old?' had everyone scratching their heads. 'Not sure,' said Willie. 'But it's been added to by many over the years, and we have done the same.' It was the biggest hut I'd stayed in so far, with a sleeping area off to the left, the middle taken up with a big dining table, and to the left a fireplace so huge I could almost walk right into it. A little alcove room provided storage for boots, gaiters, coats, hill sticks and general hill-country paraphernalia.

At the hut, we met up with Middlehurst's two shepherds, Jonty and Ben. The pair had ridden out a different way that day, mustering the river flats. Ben had come to Middlehurst straight out of school; it was his first proper high-country job. Willie knew his grandad, Lyndon Morris, and told me that Lyndon was 'an absolute legend. A great horseman who has broken in many, many horses.' Ben had inherited his grandad's knack, and rode beautifully. Back home in Cheviot, he was bringing on a young St James horse, and also hunted and showjumped. And he told me he'd spent his last pay packet on a beautiful deep mahogany saddle.

Jonty was one of those people who always wanted to be doing something. He didn't stay still for long, and so by the time we arrived the hut was warm, the firewood was sorted and even the billy was at the boil.

Over dinner, Willie shared a little tidbit of wisdom: 'You should always carry a handkerchief,' he said. 'I tie a knot in mine as a reminder.'

'A reminder of what?' I asked.

'Keep them moving' is a musterer's motto.

The George Hut is tiny in size but mighty in character. Francine Boer

'Well, anything really. I tie a knot in it, and when I pull it out later I know I need to remember something.'

'But what if you forget what that something is?' asked Grace.

'I never do,' said Willie, grinning madly.

*

Overnight, the rain came in. We heard it on the hut's tin roof, foreboding yet comforting at the same time.

So, before setting off in the morning, we donned our wet-weather kit then hotfooted it back to where we'd left off the day before. 'Things look different coming from this direction,' I said to Willie, and he nodded and said, 'That's the thing with it out here – it always looks different, and I always see it in new ways.'

The mob had spread out overnight, so the task of bringing them together started all over again. I peeled off with Ben to see if we could find any stragglers in the endless hills, spurs and gullies. 'The gullies,' he told me, 'are called the guts' – and, with bubbling little creeks running through them, I could see why. Ben was on Smit, a chunky, sportily built dark chestnut that wasn't long broken in. Ben had been working on him, taking him out in the afternoons once the station jobs were done, with Willie's encouragement. 'It's good to have someone here who is interested in the horses,' Willie said. 'And Ben is a very capable rider.'

By this time it was really raining, and little drips were trickling down my neck and off the peak of my cap. The wet made the going precarious, and as we scrambled over sheer scree hillsides Bushy Pony worked hard to keep us both safe. At one point, the ground below us started to give way and I couldn't stop my eyes from closing tight. But, of course, we were fine, and we carried on. Fear can't have too much of a hold in wild places.

From where we rode, I could see two mobs being moved along two separate ridgetops – Jonty behind one, and the girls and Willie behind the other. It made quite a sight, with the cattle stringing out into long fingers that pointed the way home.

'Carly, you can head back down and push a mob down the creek,' Ben told me. 'I'm going to see if I can find more stragglers.'

I nodded, and tried to look like I knew what I was doing.

'Just follow the creek,' Ben said, then he headed off.

Righty-ho, I thought. Simple. And I took myself off in the direction that I *hoped* was the right one because, if I was being honest with myself, I really had no idea.

It was still raining, and things were starting to feel altogether damp. I found some cows and a creek, but I lost sight of Ben, which meant I was entirely on my own with about 25 cattle to shift downstream. It was both exciting and daunting – a combination of shepherding feelings I was still coming to grips with – and the relentless rain only heightened my sense of being in the absolute middle of nowhere.

'Come on, Bushy Pony. We can do this,' I said more to myself than to my pony. She knows what she's doing even if I don't, I thought.

Our first obstacle was the creek. The cows were loath to cross it, but cross it they must. We wove our way down, and as the hours passed I just got wetter and colder. I'd been on my own for much longer than I thought I would be, and I was getting cold. Cold is not something you want to be out there, so I hopped off Bushy Pony and started walking, trying to warm myself up by jumping about. And, in that moment, I was so glad to have that mischievous pony out there with me. I'd come to trust her, and her mere presence and sure-footedness were an enormous reassurance. She'll remain in my memory as the little pony who thought big.

Eventually, and with an immense sense of relief, I spied Ben in the distance. I joined up with him, and our cows mobbed up together. We carried on pushing them down, following narrow sheep tracks for a few hours, until they blended in with the larger mob that the others had been working. When we reached a nice wide farm track, things started to feel positively civilised.

With the end of the day in sight, we shut the cattle into Mary's Block for the night, and our ride to the hut was serene. With no noisy cows in front of us, and the dogs quiet and tired, a hush fell over us all. Lulled by the rhythm of our horses, we got lost in our own thoughts. There is such a held feeling in these moments. Like, if we could just stay in it – the calm of being on a horse, the peace of what surrounds us, the shoulder-ease of a job done – then our hearts could be a more spacious and pure place.

*

George Hut was a welcome sight. An absolute classic back-country hut, so-named because of the wide river, the

Lunch break with a view. The horses take the chance to nap.
Francine Boer

Susan has many high country skills including cooking up a meal for the crew on an open fire.

George, that ran beside it. This characterful gem had a big green chimney, bright-red walls and a yellow door that tall people had to duck through. Hut perfection, long and narrow in the traditional style, it hadn't been modernised like Old Middlehurst. Susan and Willie had plans, they told me, to strip it back to its original rimu beams 'one of these days'.

Inside, it was teeny, and dinner was a feast of lamb, venison and potatoes cooked on the open fire and eaten straight from the pan. There was even a fruit cobbler and cream for pudding and, of all things, a bottle of Taittinger. 'A hut tradition,' said Susan, with a wink.

It was a good thing we had such a hearty evening, because the next day went a bit to custard. Our final-push day, time to get the 400 or so cattle back home, where Earle would be waiting to scan them. Susan was on packing-out duty, and Jonty and Ben had left earlier to ride back out to the cows. And Bushy Pony, after I had tacked her up ready for the day, decided she also wanted to be up and at it; while I ducked into the hut to grab something, she took her chance and was off, in hot pursuit of the other horses. 'You little bugger!' I muttered in disbelief when I saw her galloping away into the distance. I had no other choice but to cover those first few kilometres on foot. Eventually, a hot-faced Ben appeared, bearing a not-at-all-contrite Bushy Pony for me. I apologised profusely, feeling like a bit of a fool.

When we caught up to the boys and the mob, it was to discover that the cattle were also a bit dishevelled and unruly that morning. They must have had a group discussion during the night, because they just didn't want to move in the right direction. Everyone was a bit tired – they always are on the last day of a muster – and for some unknown reason it's often the last day when the weather turns foul. It's the nature of the work: the weather sets the tone, but you have no control over it whatsoever. This particular day we were served up drizzly rain, the kind that doesn't seem like much at first, but it just keeps going, a constant dripping tap, until it's getting on everyone's nerves. The wind, too, picked up, and soon Bushy Pony's mane was flying in all directions like an eighties shampoo commercial. The animals were as wild as the weather, but we needed to move, so the heat was put on and with red-faced hollering we trudged towards home.

Then half the mob turned on us. It's one of the worst things that can happen on a muster. All hell broke loose. Willie hooned along, trying to get the renegade cattle back, while Grace, Fran and I tried to hold the rest of the mob.

Dogs were flying, Jonty and Ben were doing some swift moves on their horses, and there was swearing and sweating. And, all the while, the clock was ticking. Ideally, the cruise home would have been plain sailing, so we could have got a start on the yard work before the day was done, but that was feeling increasingly unlikely.

Through it all, Willie kept characteristically cool, giving the orders and changing the plan every time the last one didn't quite work out. The boys joined up with us again, cattle in tow and mutterings about cows getting left behind. Not something that you want to have happen, as it means another trek out to find them and bring them back in. Another job to add to the long list.

'But sometimes things go this way,' Willie said with a shrug. 'And you just have to get on with it.'

This, after all, was the high country: a place where the going wasn't always easy, where intentions could be blown away in the wind as soon as they were laid down, where anything was possible. A grand, impressive land for grand, impressive souls.

10
WILD HORSES

St James Station, North Canterbury

When I pulled into the long tree-lined drive, I was pretty sure I was in the right place. I had spotted horses in the well laid-out paddocks, and there was the unmistakable St James brand, and when I was greeted by a man with the bow-legged stance of a long-term horse rider, I knew it. He was wearing a striped shirt, jeans, R.M. Williams boots and a large white cowboy hat. This was unmistakably Lyndon Morris, the much talked-about grandfather of Ben from Middlehurst Station. Everyone who knew him knew he was very seldom without his cowboy hat. It kind of came before him: you met it, and then you met him.

He greeted me with a cheeky gold-toothed grin, and I saw right away that, at the age of 78, Lyndon was a man who refused to slow down. 'It's a bugger getting old,' he told me as he bustled me inside for a cuppa. 'It's tough at the bottom, but the day I stop riding I may as well die.'

He wasn't kidding. He intended to keep on keeping on, and told me he still rode with a firm group of trekking friends. Just then, his wife, Barbara, came in from her sprawling garden. 'I've learnt over the years not to worry when he trundles off with a horse float hooked up to his ute,' she told me when she overheard us. 'Lynnie was born stubborn and he's a tough nut.' She left him to his horses, she told me. They were everything to him, after all, and always had been.

'When I was little I had one of those cloth books,' Lyndon told me once we were settled in together at the table. 'I was only small. I looked at a chicken, a duck and a lamb. Then I turned onto the page with a horse, and that was it. That's where my attention went. That's what I was fascinated with. Since then, I have loved horses. They are a noble animal. I love them to bits, and I have done all my life.'

His grandfather and uncle were great horsemen, he said, 'and that gene skipped over to me.' As a young boy, he wanted a pony – 'I wanted one bad' – but there was no money, and nowhere to graze one. So he hung around the

A St James horse out at Okuku Pass where Lyndon frequently rides to.

Momentos from Lyndon's rodeo and hunting days.

pony club, begging for rides. Then at age nine, he went potato-picking with his brother and diligently saved up his earnings. 'There was a pony for sale at Rangiora, and I rode my bike down there,' he said. 'I chucked the bike over the fence, got on the pony bareback, and that was it. It was the end of the bike, and I rode that pony everywhere.'

That pony was a little bay Welsh cob named Chummy, and he eventually needed shoes. That's how Lyndon met Norman Glidden. 'He put shoes on my pony, and at the same time he took me under his wing,' Lyndon recalled. Norman was the area's preferred horse-breaker, and Lyndon would go with him after school and on the weekends. 'I don't know where I would be if it wasn't for him. He taught me a lot. Ninety per cent of what I do comes from Norman.'

After several years' riding at the pony club, Lyndon decided he wanted to have a go at rodeo. 'Well,' Norman apparently said when Lyndon told him this, 'you'd better have some practice.' And he set about facilitating it. 'He got this dirty old, stinking mare out, brought her in and put a rigging on her. She had a wither on her like Mount Everest. I stepped on, and this thing came out, shot down a hundred yards and *bang!* I was slotted. I was hurt, but I got on again.

Another hundred yards and *bang!* Down I went again. I went home crawling, but I was hooked.'

Rodeo was part of Lyndon's life for 20 years, and he turned his hand to showjumping and was involved with the Brackenfield Hunt for a long stint. And when I asked him about how many horses he'd broken to saddle, he sucked in a thoughtful breath. 'Hard to say … Thousands, I'd imagine.' Many of those were St James horses.

*

The very first horse I mustered on was a St James horse. Summer, the sturdy mare who taught me the ropes and took me for that wild gallop on the heels of Alice and Kelsey up at Muller. I'd ridden a few more since.

It's fair to say that St James horses are the famous station breed of the south, and it's said that they've run in herds up the Ada Valley for well over a hundred years. With their staunchly coveted bloodlines, they hold stories and history just like the Ngahiwi, Paul Johnson and nāti horses do in the north. Nowadays, St James horses can be found all over New Zealand, and some have even been exported overseas. And, while they're still out mustering the hills, they are

St James horses enjoying fresh pasture after a long day's ride.

Lyndon in his trusty, signature cowboy hat.

also just as likely to be spotted on the hunt field or in the showjumping arena, as horses move from serving a solely practical use towards a more recreational one.

Since back in the day of horse-drawn trams and cartage wagons, St James horses have been in big demand, known for their strength and agility. In those early days, they were half-draughts — handsome, clean-legged, big types that were almost always brown — and the first sales of them drew in buyers looking to add to their large horse teams in farms throughout Canterbury. There was much pride in owning a St James horse.

The St James sales were held every couple of years, and getting the horses out of the Ada Valley — where they roam free, ranging over the Waiau River lands — and into the saleyards had always been done by riders on horseback. It was a 50-kilometre muster, ending at the St James Homestead yards. The horses were pretty close to being wild, and getting them in was not an easy task, but I'd heard the St James muster was an adrenaline-filled adventure for those who were brave enough to tackle it. The stuff of legends.

*

The Last Muster

Lyndon had been on 17 St James sale musters, and twice that number of gelding musters. He still had the note written by Jim Stevenson, the founder of the St James breed, on a blue piece of paper inviting him on his first muster back in the eighties. 'It's very hard to break into the muster,' Lyndon said. 'A hundred have tried and they have failed. It was a wild one back then, and I loved it.'

Lyndon did his last St James run in 2022, at the age of 77, but the muster still lived strongly in his memory. 'You had to keep up with the young horses, and a horse without a saddle on is always going to be quicker. Every time. We would drive them up, and they would break, and we would just have to start again. They would always want to go back to the Ada.'

There was one particular muster, he said, where the horses kept turning on them. 'Five times, we musterers had to go back in for them. Five long times. Everyone was getting sore and tired, and they were about to give up and cancel the sale. So I said to Jim, "Give me one more crack at these horses.' And he said, "Okay." Dave Ferriman, a good huntsman, came with me, and we had a stroke of luck. We got them in all right! It was the longest, toughest day. But we got the buggers.'

There was another muster that went smoother, and the team got the horses in the yards early, 'but we had lost one here and one there,' Lyndon told me. 'And we had five days up our sleeve, and so I said, "Let's go back in."' So, rather than congratulating themselves and having a beer, Lyndon and Dave turned tail and rode back into the Ada and camped the night. 'We found four horses up there, and so away we went. We picked up another one at Strawberry Gully, and we were going pretty good. Then, heading down the Clarence, we picked up another one on the flat.' They rode back out with a string of seven horses and a swagger they had earnt.

Lyndon gave one of his wry chuckles, noting that it was the only time he recalled Jim Stevenson giving him a nod 'and a sort of smile'. 'Jeez, that man was tough, and he wasn't one for compliments. He could ride any horse in his day. He wasn't tall, but he would sit on a horse straight up and down. He could ride, all right.'

Nerves of steel is what you needed on those musters, Lyndon told me – and, I thought, a certain amount of bravado and stubbornness, too. He'd often get given a hot-headed bucking horse to ride. 'I'd get slotted on the naughty one, because I guess Jim thought I could ride.' That he most

Some good old musterer's fayre. It's never fancy but it's just what is needed after a long day in the saddle.

Lyndon Morris has clocked up some hefty hours and miles in his saddle and he favours a St James horse.

certainly could: riding horses, backing them and starting them was his superpower.

Lyndon told me he still had a few tricks up his sleeve. He was determined to pass on what he'd learnt from his many years' mustering and breaking in 'to the right kind of people' – just as Norman had done for him. One of those people was Sophie Logan, a young horsewoman who had brought a horse to him for breaking in 2017. She'd got involved in the process, absorbing everything, and when it came time for her to take the horse home, Lyndon asked if she wanted a job. He'd seen a certain something in her. 'She has a knack, and we get on good,' he said. 'And she is tough, all right. She has an affinity with horses.'

So, while finishing off her Bachelor of Science majoring in biology and psychology, Sophie had carried on working alongside Lyndon and the horses when she could. And, thanks to his confidence in her, she ended up starting her own business breaking in and handling horses. Lyndon might have stopped putting up his hand to take on the young St James horses that owners needed handling after the sale, but now someone like Sophie could do it instead.

In his long life with horses at his side, Lyndon had also been a shepherd. Naturally. He went on to manage then part-own Okuku Pass Station, in Hurunui in North Canterbury. And, through it all, always, he used horses. 'It's the only way to go. On a horse, with your dogs in behind you. Best thing in the world,' he said. 'It has to be horses. The stock move better. It just works. It's lovely and so relaxing. Your horse is your friend. There's nothing better than that.'

*

Nothing much deters horse people from their task once their minds are set, least of all a bit of drizzle. As I headed up to the St James Conservation Area one grey Saturday, I followed a procession of four-wheel drives on a road that had been graded in anticipation of the sale, but the overnight rain had turned the surface into tea-coloured sludge. There was a view out there, but I couldn't see it, thanks to the heavy fog that had moved in.

It had been a few years since the last sale, and as I pulled up I could see the turnout was big. There was a definite air of excitement. The St James horse sale wasn't just about the horses; it was also about high-country catch-ups and farmy chin-wags. Standing alongside the high wooden yards, I

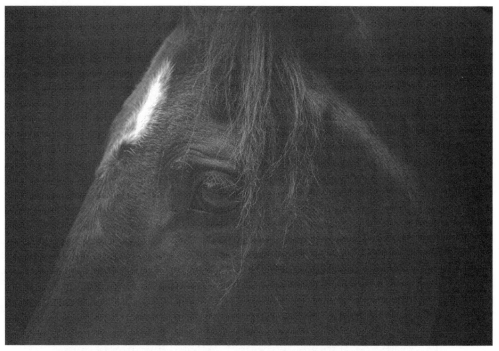

St James horses live their first years in the hills of the Ada Valley and the sale is their first introduction to domestication. Ava Thomas

The horses are very nearly wild and this new environment can be frightening but they adjust quickly. Ava Thomas

caught little snippets of how that year's muster had gone. It sounded like a tough one, and when Ben Dampier-Crossley got up with the microphone to welcome everyone, he confirmed this. 'We were out there a bit longer this time than was expected, after some of the horses decided they wanted to go back to their home territory,' he drawled, 'but the guys who come out to muster are never too worried about having to stay out there.'

Ben was the grandson of Jim Stevenson, and this land was owned by the Dampier-Crossley family for 82 years. In 2009, the station was bought by the government and now managed by the Department of Conservation as the 78,000-hectare St James Conservation Area. Since then, the Dampier-Crossley family had continued to run horses there, keeping the tradition alive. After doing a shepherding stint at St James in the 1970s, Peter Newton had written that Jim Stevenson was 'quiet and unassuming … one of those blokes that doesn't sidestep any job. He's the sort of man for that sort of country.' And that sort of country was far from easy. It ran along the Clarence River with the Captain Range, from Jack's Pass to the head of Stony Creek, as its front boundary. From there, it reached all the way through to the Lewis Pass and the 1140-metre peak of Mount Belvedere, and beyond that to the Main Divide. Massive country that took great tenacity and courage.

And while much of the station was now conservation land – spanning three mountain ranges where tramping and hunting were the main activities – the horses in the Ada Valley had ensured that Jim's legacy lived on. Still today, the breed remained part of New Zealand's equine and mustering history.

*

The St James horses I saw in the ring that year weren't as heavy-set as the original ones. Times had changed, and with it the horses' workloads: they'd gone from pulling wagons to carrying riders across all sorts of terrain and disciplines. The 20 geldings and fillies were by four different stallions, and there was everything from tall, high-stepping youngsters to more compact, gutsy types. But it was the calm way they all navigated their strange surroundings that stood out to the crowd gawping at them. These horses had, just days before, been running as a wild herd in vast country, and now their new lives were being decided by potential buyers.

The auctioneers take charge with their fast banter. Ava Thomas

The horses are encouraged to move around the yard so that potential buyers can see their paces. Ava Thomas

The musterers were in the yards, pushing the horses through a few at a time, so the crowd could get a decent look at the way they moved. People gathered, ticking their lists and making notes. It was a great place for people watching, with a combination of old-timers in cowboy hats and oilskin coats, station owners scoping out their next break-in project and horsey people from all disciplines. You could pick the showjumpers by the sponsorship patches on their jackets and caps, the musterers by their Wrangler jeans and battered boots, and the trekkers were the ones who had dressed for the colder weather.

I spotted the Muller Station crew, and asked Alice which one was her pick. We both agreed that number 12 was nicely put together.

The crowd mobbed up like a perfectly mustered head of cattle, and we all settled in, eager to see who was buying. A woman in a pink cowgirl hat was the first, and after that the bidding got into full swing. The hushed crowd, the crackling microphone, the fast-talking auctioneer and the hum of excitement. The horses were selling at between $3000 and $5000, and I suppressed a ridiculous urge to raise my hand.

When it came to number 12, the bidding became a little more fierce. The price rose higher and higher. 'Sold to

Muller Station!' the auctioneer finally called. 'That filly has found a good home out there.' Another horse, number 17, also went to Muller. A couple more St James horses to add to the station's mustering crew. It was nice to know that at least two of the horses would be returning to the hills.

11
FREE REIN

Bluff Station, Rangiauria/Pitt Island, Rēkohu/Chatham Islands

Getting to Rangiauria/Pitt Island, an outlying rock in the middle of the Pacific Ocean, is not the easiest thing to do. The stepping stone to Rangiauria is Rēkohu, the main island, 863 kilometres east of Ōtautahi/Christchurch, over a wide expanse of sea and clouds.

With my saddle squished into my bulging bag I took the two-hour flight to Rēkohu, then bunked down with my mate Jacqui and her family, island locals who lived in the teeniest tiniest fishing settlement of Owenga. 'There's probably more horses than people over on Pitt,' Jacqui told me. 'The Bluff Station cattle muster is a wild time, I'll warn you now.'

But, for a while I just had to wait, plotting my next move over to the island where only 40 inhabitants lived. First, I thought I might be able to jump aboard a boat full of musterers who were meant to be making the three-hour journey over, but a birthday party that lasted three days scuppered that option. Then, there was the plane – a toy-sized five-seater that flew on demand. I narrowed my sights on that. I mucked about on the island for a few days, and met Freya, a gorgeous city-slicker from Northland who had switched to being a shepherd in a mid-life turnaround. She was heading over for the muster, too, so we became fast friends, joining forces while we waited to make that last hurdle over the sea.

Finally, the call came in: *Be ready, the plane is winging its way over to Pitt to pick up some tourists and you can jump in.* We met John the pilot, shook his hand, and at last we were making our way into the air and over the Pitt Strait. Soon, the dreamed-of destination loomed large below us. It was much greener than Rēkohu, with one road snaking through the centre. Houses were far and few, and the white tips of waves crashed onto the hips of rocks. We landed in a paddock and into an adventure that was just waiting to begin.

*

The muster wasn't starting for a few more days, so we had more time to kill. We spent it hanging out with seventh-generation Pitt Islander Jamie and his 12-year-old daughter Paige, who didn't seem to mind having us dossing in her room. There was, fortunately for us, a real open-door hospitality here on the island.

Jamie was a fisherman, and a good horseman too, so Freya and I soon found ourselves saddling up two of his horses and cantering along the vast expanse of the island, bashing our way through harekeke and tarahinau. Jamie was up on his freshly broken quarter horse, Buck, and Paige was on her trusty mare, Galaxy. As we trundled along, the pair gave us the ins and outs of the island.

'Basically, everyone is related or connected in some way,' said Jamie, before explaining that Frederick and Mary Hunt were early European settlers there, and his own family dated back to them. Many of the people living on Pitt had a branch of their tree that connected to those early pioneering farmers, and so did the sheep – Frederick introduced the first herd there in 1842.

The windswept island was a bell-jar community as tightly knit as a handspun jumper, and here the kids were free to roam, knowing they could go to any house and be fed. Horses, too, had a free rein. 'This is horse country,' said Jamie. 'There are motorbikes now, but not that long ago the only way to get about was with a horse.'

I asked him about Bluff Station, and the Moffetts, its owners. 'They still live very much the old way,' Jamie said. 'Moff is an amazing horseman and when you step into Bluff, you'll be stepping back in time.'

*

James Moffett, fondly known to the locals as Moff, dropped his chin modestly when I mentioned his legendary status to him the next day. 'Ah, no. I've just picked up things from here and there,' he said. 'I just love horses, always have.'

Moff was a big man. Everything about him was big – his hands, his barrel chest, his smile and his land. Six thousand acres was a big spread on a small island. He sat at the table like a gentle king, completely at ease with not just his surroundings but his place in the world.

The Bluff Station homestead sat up the hill from Flowerpot Bay – the main bay of the island, where fishing boats were launched from the wharf – and the big old two-storey home had no pretensions. This was where Moff's daughter Eileen and her three kids lived, along with chickens running wild, a rambling orchard and the Pacific Ocean as the nearest neighbour. There was no mains power on this little island, and the Bluff house used fire, diesel and gas to cook with. There was a generator for lights in the evenings.

Eileen had milked the cow that morning, gathered wood for the fire and baked the slice we were now nibbling on. She had been brought up in this very house, and was an epic horsewoman. After working in shearing gangs, then at Waitangi West station on Rēkohu, she had moved back to the homestead with her sons Bowen, Buck and Reuben in 2020. This single mum was an integral part of the station and, curled up like a cat in her favourite chair by the fire, she quietly plotted a muster plan with her dad. There was a cattle beast to be killed for eating, the cook house needed ship-shaping, and it was still unclear when those party-goers would be arriving …

Tussock, the station's right-hand man, pulled up a seat at the table, and reached for the teapot. He had come over to fish, then farm on Pitt about 15 years ago, and, 'Well, I fell in love with this place,' he told me, running a hand through his grey beard. 'The Chathams is good, but here on Pitt, well, it's the best. I love the hills, and I can sit and look over at a gully and see pigs to hunt, and that's the life I want.' A quietly spoken man who made for easy company, Tussock lived at the station's old shearers' quarters and was as good as family. He'd been brought up on horseback in the back blocks of the Manawatū, and his horse, Boy, was his best mate.

'So what time do you guys usually start work in the morning?' I asked, and got a burst of laughter as a response.

'Everything takes its own sweet time here,' said Tussock.

'There is never any rush,' added Moff.

*

'We'd better go find some horses, I suppose,' said Moff later that afternoon. The musterers were still stuck on Rēkohu, so a horse rounding-up mission had been plotted.

The island was precariously close to the Roaring Forties, where black clouds could scuttle in within minutes, and the weather had just moved in as I jumped on the back of Tussock's quad bike with seven-year-old Bowen. As we followed Moff and 19-year-old Reuben through the

Pitt Island from the air. Our first glimpse of our island home for the next few weeks.

The Pitt Island population swells as friends and family come over for the muster. It's a highlight of the year.

encroaching cloud, I caught glimpses of the rugged coastline. It was starkly beautiful and fantastically foreboding.

The herd turned tail when we found them, and the big grey heavy-set horses that Moff was known for cantered fast before skidding to a huffing-and-puffing halt. The thunder rolled in, lightning lit up the scene and the heavens opened just as we got the horses into a yard.

The bloodlines of these horses, with their manes hanging low down their broad shoulders, were old. Moff's father, Jim, had run a draught stallion and the horses' genealogy – just like that of Pitt's human inhabitants – was connected to the island. As a young man, Jim had come over to manage the farm, working hard and eventually buying the land and integrating himself into the island. He was what the locals called 'an import' – the word for anyone not tied to those original settler families – but the Moffetts were very much an accepted and respected Pitt family.

The biggest horse of them all was standing right in front of me. A beautiful dapple-grey, majestic and mammoth. I looked like a Shetland pony in comparison. This horse was one that Reuben had been tasked to start; it was his turn to learn the generational skill of making a good station horse. But, right then, the rain was coming down hard, so we tied the horses up to wind-whipped kōpī trees, and high-tailed it to Moff's house to wait out the storm.

The house was on the east side of the family land, and Moff had built it himself out of timber milled from the farm, then he had lined it with the very earth it stood on. It was curved, quirky and surrounded by bush. The outside toilet had a wall made out of coloured glass bottles, and little arty touches had been added in here and there. There was no road to Moff's house, and access was not easy, especially in winter when the clay-based soil turned to cloggy mud. If you wanted to visit Moff after the rains had set in, you'd better be able to ride a horse. On the wide windowsills of the house were treasures offered up by the sea, and below was Tupuangi, an expanse of white sand and eternally crashing waves. This was isolated living at its best, and it suited Moff just fine.

Surrounded by buckets of fruit gathered from his orchard, jars of homemade preserves and flagons of dubious-looking spirits, we settled in while the storm knocked at the walls and I asked Moff about the muster. 'It's smaller now,' he said. 'It was much bigger when we used to have the land over at Glory Bay.'

Glory Bay was at the far end of the island, and was named after the 100-tonne sealing brig *Glory* that was wrecked there in 1827. The tale goes that a nearby schooner, the *Samuel*, helped pull *Glory* off the rock she was stuck on, only to then have her founder on what is now known as the Glory Reef. The crew salvaged what they could, made it ashore and, using one of the ship's still-intact small boats, sailed and rowed to Kororāreka in the Bay of Islands. It went down in history as one of New Zealand's most epic maritime survival stories, and the *Glory*'s anchor now leans against the old shepherds' hut in the bay. Glory Cottage, a straight-up-and-down weatherboard beauty, was built in the early 1860s, and when the Moffetts farmed the area they used it as their base during musters.

'One year at Glory,' remembered Tussock, 'we marked 1800 calves in three days. And we went through 24 litres of whiskey in 24 hours.'

Reuben's eyes widened, and Bowen whacked his grandad with a cushion. 'Naughty Grandad!' he admonished.

'Ah, it's all about enjoying it,' said Moff, chuckling softly and pulling his grandson into a rough hug.

The station just ran cattle now. When wool prices reached an all-time low in the nineties, the sheep were shipped off and the red shearing shed became a landmark on the hill pointing visitors to Bluff homestead's rickety track. Pitt Island cattle were known to be hardy, and when they turned up at the stockyards in Timaru, buyers lined up. Devon crosses, the Moffett cattle were quiet – and that came down to the renowned way in which they were handled. 'Using horses to work with cattle is the best way there is,' said Moff. 'It's all about educating them to move slowly. Horses are essential here – I'll always take a horse over a bike. Without them, I'd be broke.'

Nothing is easy when you live on a tiny, far-flung island, and getting cattle to the mainland could be a mind-blowingly difficult task. For starters, there was only one ship equipped for the job, and it was intermittent and heavily weather dependent. Pitt farmers were completely at its mercy. Winters were tough, and if the ship couldn't come to get animals off, then stock had to be killed rather than letting them starve to death. Earlier, Eileen had told me they could never farm to the capabilities of the land. Instead, she said, 'We have to farm to the capabilities of the ship.'

And then there was the job of getting the cattle onto the ship when it did arrive. First they had to be mustered to the wharf, next they were put on a barge, and finally they were taken out to the ship bound for New Zealand. Quiet cattle were a necessity, so the Moffetts' approach to mustering them was a bit different.

'You'll see what we mean tomorrow,' promised Tussock.

*

The next day we were finally saddling up our horses. All the gear here was about practicality rather than looks: old leather saddles were well worn, breastplates were fashioned from mismatched leather, and bridles were made from good old green cray rope. Used to make crayfish pots, cray rope was the number-one fix-it tool here on Bluff Station. Gates were hinged and tied with it, and Moff's pants were held up by it. Nothing was showy, everything had a purpose and my crappy old saddle finally fitted in.

DD, the station's laid-back stock manager, offered me a hongi and a hug when he met me that morning. He was an East Coaster born and bred, a Ngāti Porou gem, and I immediately warmed to his easy and welcoming manner. He lived down the hill, in a little house called Sleepy Hollow, with his wife, Ashley, who was pregnant, their four sons and their extended whānau.

The whisper of mustering action had been blown around the island on the ever-present Pitt wind, and more and more locals congregated. There was a buzz in the air, and Moff was in high spirits. 'He loves this,' his son Mike said to me, as he threw some steaks onto the sizzling base of an ancient whale pot. Broad and strong like his dad, Mike was a traveller, seeking a world bigger than the little island he'd been raised on – but he returned for the muster whenever he could. 'Having everyone here, having a good feed. This is what it's about.' Mike grinned.

The cattle beast had been killed the day before, and the men had gathered, working together, drinking beer and joking. This muster was the highlight of their year, an important social time on an island that could be a lonely place. Eileen said it was a chance for the family to celebrate the station with everyone, to connect with the community and to show some island hospitality. And it was finally getting started.

I climbed up on Suzie, an ex-pacer mare that had been given to DD. Pacers were a common sight on Rēkohu where, surprisingly, there was a racecourse. The oldest one in New Zealand, in fact – it dates back to 1873, and was a big part of island life, as were the horses themselves. Just a few years

Author Carly Thomas on Pete, one of the Moffett's big grey horses bred on the island. Eileen Moffett

The way of mustering on the island is very laid back and quiet. Things are done without stress.

earlier, in 1867, there had been about a thousand horses on Rēkohu. Ngāti Mutunga o Wharekauri, the main Māori iwi of the time, traded potato crops for breeding mares, mainly thoroughbreds, and those original bloodlines still lingered on the bigger island. The pacers, imported from New Zealand, were generally ones that had been too slow for racing on the mainland; and, when their island racing careers ended, some were brought over to Pitt, adding in some lighter offspring to the heavy-set Bluff stallion. Riley, a teenager who was filling in her school holidays on the muster, was also on a pacer.

'You look a picture, Sal!' I called out to Sally, a big-hearted Australian with a golden sense of humour who I'd met that morning. And she really did – blonde hair streaming, a *Man from Snowy River* oilskin coat skimming her ankles and a bottle of beer in her hand. She laughed. 'I look like something, that's for sure.' She had lived on Pitt for a stint, and was back for the muster, up on Bob, one of the big grey Moffett horses. Bob was absolutely huge. I'd been told he loved to work cattle, but that he was also known for his blind bolt.

Nearby, I spotted Moff, up on his trusty horse Not So. The horses all had names that came with a story. Not So was, Moff explained, 'not so fast and not so slow'.

Reuben rode up beside me. 'If you ride a horse for the first time here, you get to name it,' he told me. 'DD's horse has two names put together, because two people claim to have ridden it first.' Concrete Lady was therefore the dappled grey's unlikely name. She still had a foal feeding from her, so the feisty youngster also tagged along on the muster.

'So what's your horse's name, Reuben?' I asked.

'Seat Eater,' he called over his shoulder, as his horse skittered round in a circle. 'He has a thing for eating bike seats.'

One of Ashley's nephews, Paku, was there too, and had a little speaker tied to the pommel of his western saddle. It was already pumping the music that would become a theme of this muster. The teenagers stuck together, all on pacers, heads bobbing along to hip-hop beats, riding with that ease I'd come to know from East Coasters. The rest of us would join in on the chorus when something older and more familiar came on – Fleetwood Mac, Johnny Cash, Crowded House. The soundtrack was random, just like our motley crew of musterers.

Moff's other son, James, led the way on a big bay with a wide white blaze. His horse was called Slicker. 'We had another horse called Slick,' Moff told me, 'but this one is Slicker than Slick.'

We headed out to the coastline, the edge of the island that had a magnetic pull. Spirits were high, and they were also being passed around: strange, multicoloured alcoholic concoctions dubbed shepherds' whiskey that made you cringe first, then giggle. Sips of home-brewed blackberry gin and swigs of port from hip flasks warmed our bellies, and the horses pranced along, picking up on the excitement. This muster, I could already see, was really just a thinly veiled party that lasted for days.

Through vast open paddocks we rode. The dogs, running like snakes in all directions, were wonderfully dishevelled and happy. Moff liked his bearded collies, a shaggier version of the dogs I was used to seeing out mustering. Not as loud and bolshie as huntaways or quite as cunning as heading dogs, they were somewhere in between. Their laid-back nature suited this place.

We dropped down to the bone-white sands of Tupuangi. Waves crashed on a reef and tōrea pango and tuturuata scurried along the shore. It was incredibly beautiful, and just knowing that this stretch of beach could go for days, weeks or months without having a human upon it made it feel that much wilder.

The history of the Chatham Islands is layered with peace and conflict, and on Pitt those echoes sat close to the surface. The Moriori were this island's first inhabitants, and lived here for at least 500 years before Europeans ever arrived. They were peaceful hunter-gatherers who, through their isolation, developed their own customs and dialect distinct from the iwi of Aotearoa. They had outlawed war and killing, and on this sweep of sand it was not hard to imagine those times of peace when Moriori lived in idyllic isolation.

In 1791 an English ship, the *Chatham*, was blown off course and found the main island. Later, transient hunters of sealskins and whales began to visit bringing change and disease. Then, in 1835 two invading Māori groups brought battle and conflict. Death and enslavement followed, and the Moriori, who decided not to fight, suffered extreme loss. There was a heaviness to the Chathams and the ancestral pain whispered in the waves and the crevices of the hills.

Suzie skittered at the edge of the waves, and Tussock told me that great white sharks lived there and were respected by Pitt Islanders. 'Swimming is not really a thing here,' he said.

Then he told me to look out for whale vomit.

'Whale what?' I asked.

'Ambergris,' he replied. 'They use it in perfumes, and it can be worth a bit.'

And so I scanned the ground for a rock that I was told would look a bit different from the rest. Anything seemed possible in this untouched place.

*

Off the beach and into the bush, crooked kōpī trees closed in on us as we climbed towards the base of towering Mount Hakepa, or Walk-'em-up as the locals called it. With cattle finally in our sights, and DD flushing them out of the bush above us, we started to peel off in different directions. I headed high with James, Sal and Riley, and from our wind-blown vantage point we could see Jamie, Paige and Freya in the distance, little dots moving closer, having ridden over from their corner of the island. Quad bikes were also closing in on the lower reaches, and with horses and people in all directions it looked like we were part of some mad military operation.

We had cattle in front of us now, and as they mobbed up I started to see the different way in which the Moffetts worked their stock. Tussock would take the lead and I could see the cattle were used to following him and his steady grey horse. With horses and riders stretched out behind and around the mob, we moved them as a whole. Moff urged everyone to give them space, keeping them calm and not stressed, and rather than the cattle breaking away we would take breaks often. The motley bottles of shepherds' whiskey were passed around, legs were stretched and the mums and calves were given a breather. Nothing hurried, no care in the world as to what time it was.

'Just enjoy yourself,' Moff said more than once.

Up the hill and through a wire gate and that was it. We were leaving the cattle there until the next day. They were remarkably unstressed. This slow way was certainly easy on the cattle.

*

Later that night, we feasted on freshly caught blue cod dipped in beer batter and fried in beef fat on the whale pot. It was the best thing I had ever tasted. So good I was almost in tears. Kaimoana was abundant here, and as the saying goes, if you're hungry on the Chatham Islands then there's something wrong with you.

Moff came and sat next to me on the wobbly plank seat, and I took the chance to ask him about his technique with young horses. 'I use the Parelli approach,' he said. 'There was a clinic held over here years ago. It blew me away, and I went towards it. It's a gentle way, but at the same time you don't let them get away with anything. It gives you tools so you can teach them at every stage.'

Reuben planted himself next to his grandad. 'The main reason Grandad uses that technique is because he got sick of running around the paddock chasing his horses,' he said.

They grinned at each other.

'You want the horse to want to do things,' said Moff. 'It doesn't have to be a battle.'

*

A shower of rain overnight, carried in from ocean clouds, turned the yards into a mud bath. We spent the morning draughting the calves and ear-marking them, and were soon filthy. The men were pushing the cattle up, filling the pens and making a game out of it, showing off, jostling each other and being as raucous as possible. The music was cranked, and the occasional bursts of singing and dancing added to the party atmosphere.

Leaning on the rails was Chatham Islander Travis King, who told me he'd worked for Lyndon Morris many moons ago. 'Lynnie taught me a lot of what I know now,' he said. It was an instance of the wonderful serendipity offered up by the close-knit horse community.

Lynnie would have loved this, I thought as I saddled up big old Bob so I could help take the cattle back out to the paddock. When I'd begged to ride on a Moffett-bred horse, Sal had passed me Bob's reins. We made a funny pair, me being smaller than most and him being the biggest horse on Pitt. He was certainly powerful, and as we cantered to cut off some runaway calves I momentarily let him be in charge. Bob knew what he was doing, and if he could have talked I'm sure he would have told me to just hang on.

*

Unlike any other muster I had been on, we took days off on the Pitt muster to go fishing and catch up on island chores. I carted firewood, baked cakes, went for wild walks with the

James Moffett keeps the cattle mobbed up and moving towards home.

Slowly, slowly the large mob of cattle gathered from all the corners of the station are almost home.

kids, and Bowen pointed out the island's birds to me. 'Aren't they so beautiful?' he said. 'Look at their feathers, Carly!'

Sal also took me for a jaunt to meet Aunty Di, the island's oldest resident. She had spent 61 years on the island and told me, with her little dog sitting on her knee, how horses used to be the only transport on Pitt. 'They pulled a cart to pick up supplies from the boat from New Zealand. It only came once a year back then. And the kids rode horses to school in whatever weather.' She leant towards me, finger pointed to emphasise that she wanted me to understand what she was saying. 'Horses were a tool, and they got us where we needed to go. We didn't need anything else.'

*

The mustering crew from Rēkohu, the ones who had been waylaid by that epic party, finally arrived by boat at the Pitt wharf. The islanders referred to the wharf in the bay below Bluff as 'town', although it was really just a trio of fishing boats, a cattle yard, a school and a historic church.

The boatload of rogue musterers rocked up at Bluff with booze, lollies, the biggest Bluetooth speaker I had ever seen, and a fresh collection of lungs for singing, laughing and yarning. Brothers Claude and TC had been doing this muster for many moons.

'It's a brotherhood,' declared TC.

'The best place to be,' said Claude.

TC had brought his little and ancient dog over on the boat. Gizzy the terrier was 15 years old, and had been over for 15 musters. 'He actually died last time,' said Claude with such a straight face I realised he wasn't joking. 'But TC resuscitated him, and here he is.' The old dog still sat up front on TC's motorbike, and although he couldn't see or hear much, he still did his bit and barked hoarsely at the cattle.

The population of the island had expanded hugely, and I found a quiet moment in the homestead kitchen with Eileen, who calmly kept everything ticking along. She baked six loaves of bread a day, made sure there was always something to eat and wrangled the kids while still managing to jump on a horse from time to time. She also had the most incredible garden, its neatly tilled rows bursting with vegetables. 'She has a knack,' Tussock had murmured to me one evening over a dinner of roast mutton and freshly dug carrots that Eileen had prepared.

'Living out here you have to be able to do everything,' Eileen told me simply.

The Last Muster

*

And so it was with a big crew that we rode the horses out to Waipāua the next day. It was a decent journey, so we planned to spend the whole day riding the horses over. That way, we could muster the cattle homeward the following day.

The sun had risen to the occasion, the wind had dropped, and as we rode along the now familiar Tupuangi, the sea shone a crystalline blue. Reuben was up on Jughead – yes, he had named the horse with a huge head, and was starting it with the aid of his grandad. Additionally, Reuben hadn't long transitioned to a saddle; in the Moffett family, it was tradition to ride bareback until you were 18. 'It means they get a good seat,' said Moff. 'And they rely on riding skills rather than gear.'

As I rode alongside Reuben on the big grey gelding and Moff on Not So, it was neat to see their techniques for encouraging forward movement. Moff used his horse, and Reuben used his body to keep the novice horse on track. The emphasis was on forming a partnership, and I could see that transpiring – not just between Reuben and his horse, but between him and his grandad, too.

*

The next day of mustering was my biggest yet on Pitt, and by the end of it we had a big mob of cattle strung in a long procession along the one and only road. DD had taken me up to the front of the mob. 'It's where the action is!' he'd said as I cantered along behind him. And indeed it was.

When the going opened out, we had to make sure the cattle moved in the right direction and didn't break away. We moved in unison with the mob rather than battling them, and this style of mustering seemed to tie in with Moff's ethos: 'Simple is best, and there's no hurry. Life is too good to hurry through.'

As Moff explained, 'Here there's no need to make things hard. We don't need much. We have good things in abundance, so why not enjoy it all?'

*

My days on Pitt had turned into weeks, and with the cattle finally making their slow way into the Bluff yards, my mind started to turn towards home and how I was going to get

The Last Muster

back there. But then the clouds rolled in, bringing the kind of weather that wasn't suited to tiny planes.

I needed no prompting to saddle up and help with taking the freshly ear-marked calves back out to their grazing, and I was only too glad to help out on the final straggle muster, collecting up all the cattle that had escaped the main musters.

One day turned into a few more, and I became like an islander, gazing up at the clouds and giving unfounded opinions on when they would part. Moff told me, in his characteristically philosophical way, that it was all meant to be, and I had to agree with him when I found myself closing the gate on the last mob gathered.

'That was my last cow, DD,' I said, as the gate latch clunked closed, twisting around in my saddle to see him. 'And my last muster.'

'Well,' he said, with one hand extended and the other over his heart, 'it's been an honour having you here.'

*

Just like that, my mustering adventure was over. There wasn't a backslap of congratulations. No big, conclusive, celebratory moment. There was something much more satisfying: a ride home in the company of people who knew the worth of a good horse.

I had wanted to understand more about this way of life, to rub shoulders with those who used horses to do their mustering work. And I'd done just that, finding in the process a shared thread of true passion and determination that rested in the hearts of the dedicated characters on dusty back-country roads who loved horses.

From the top of the motu to the bottom, and on the islands beyond, every place had opened their doors to me. They had said yes to me just as they had said yes to horses. The mighty, steady animals that had carried me up and over, down and through. So many horses – St James, Ngahiwi, standardbreds, quarter horses, the huge Clydesdale crosses and a few that were a mixture of many. They had all given me a better understanding of their worth and their why.

Mustering on horses, I had learnt, was a simple act, but one that was honest. It was layered with history, and chosen by those who sought out an older way, a slower way. A good way. Because, as Moff told me when I sat down that night on an upturned bucket with a cup of tea in my hand, 'Life doesn't have to be pushed, and it doesn't have to be hurried. Horses can do that for us, and what more could you need?'

I cannot believe that this way of life will ever be lost completely, not while people like all those I've had the pleasure and privilege of riding with still know the worth of mustering on horseback. Not while they are still actively passing the skill on to the next generation. Not while it works, on a practical and on a spiritual level.

High-country ways are being elbowed in on – I saw that along the way – but the people I met were tenacious. They, and those like them, adapt. And where they go, so do their horses. Yes, there will be more last musters, just as on Greenstone, but at the heart of it all – at the centre of this coiled rope of knowledge – burns a love for horses that is hard-won and tightly knotted. Horses carry us home, and to ride them is to know adventure pure and true.

ACKNOWLEDGEMENTS

Thanks to my family who kept things ticking, wrangled the laundry mountain while I was adventuring and never once made me feel guilty about continually waving goodbye during an entire year.

Thanks to all the people that said yes, the ones who welcomed me on their musters after just one phone call, offered me a bed, food, a horse and country hospitality.

To those who went a step further – The Stevenson family at Upcot Station – a family I could turn to for a bit of home comfort when I really needed it and for just being epic people. And Bruce and Karen at Karetu Downs, thanks for your generosity, I got so much done in your cottage.

Thanks to my brainstorming, connected and always supportive friend Lindy Relling who listened, pointed me in the right direction and opened many doors for me.

A shout-out to all the neat friends and strangers I chatted to along the way. You helped me write this book by listening to my stories and letting me form them into words. I love a good natter and you made my day when you stopped to chat.

Hugs for gorgeous Fran and my clever daughter Ava for helping with photos.

And, lastly, gratitude for all the horses who carried me and helped me to solidify my heartfelt knowledge that horses have helped me at every turn.